THE BATTLE FOR
HUMAN NATURE
Science, Morality and
Modern Life

BOOKS BY BARRY SCHWARTZ

Psychology of Learning and Behavior
Psychology of Learning: Readings in Behavior Theory (ed.)
Behaviorism, Science, and Human Nature (with Hugh Lacey)

Barry Schwartz

THE BATTLE FOR HUMAN NATURE
Science, Morality and Modern Life

W. W. NORTON & COMPANY

New York • London

Published simultaneously in Canada by Penguin Books Canada Ltd, 2801 John Street,
Markham, Ontario L3R 1B4
Printed in the United States of America.

The text of this book is composed in Fairfield, with
display type set in Century Nova. Composition and
manufacturing by the Maple-Vail Book Manufacturing Group.
Book design by Nancy Dale Muldoon.

First Edition

Library of Congress Cataloging-in-Publication Data
Schwartz, Barry, 1946–
   The battle for human nature.
   Includes index.
   1. Sociobiology.   2. Economics.   I. Title.
GN365.9.S38   1986   304.5   85–32057

ISBN 0-393-02319-2

W. W. Norton & Company, Inc., 500 Fifth Avenue, New York, N.Y. 10110
W. W. Norton & Company Ltd., 37 Great Russell Street, London WC1B 3NU

1 2 3 4 5 6 7 8 9 0

For RACHMIEL BEN PINCHAS v'SHAYNA SARA, who painstakingly
  showed me the problem.
And for MYRNA, who daily shows me its solution.

130235

Contents

Acknowledgments

The inspiration to begin the studies that resulted in this book grew out of a series of formal and informal collaborations with several colleagues at Swarthmore College. Swarthmore is one of the few institutions where the pursuit of understanding is allowed to cross disciplinary boundaries, and I am indebted to Ken Sharpe, David Weiman, Braulio Munoz, Jim Kurth, Thom Bradley, Hugh Lacey, and Richard Schuldenfrei for guiding me in that pursuit. I owe a special debt to these last two, with whom I have talked, taught, and written almost daily over the last ten years.

Many people have suffered through early versions of the book and provided criticisms and advice that have, I hope, improved it. While I will be happy to accept whatever praise comes this way, blame should be directed at the following: David Apfel, Jon Baron, Mark Breibart, Charles Dyke, Dick Herrnstein, Alan Heubert, Scott Gilbert, Herb Gintis, Hugh Lacey, Morris Moscovitch, Hans Oberdeik, Dan Reisberg, Rob Richards, Paul Rozin, Jon Schull, Richard Schuldenfrei, Myrna Schwartz, Marty Seligman, Ken Sharpe, Alan Silberberg and David Weiman. Norton editor Don Lamm should be held especially culpable, since he had the last chance to correct errors of style and substance.

Reading and writing take time, and I am grateful to the National Science Foundation for providing me with a decade of research support that helped make the time available. I also could never have gotten started wthout the help of a Eugene M. Lang Fellowship that supported the sabbatical leave in which this book began. I am also indebted to Didi Beebe, whose prompt and skilled secretarial assistance allowed me to focus all of my energy on matters of substance.

I am thankful to all of the above individuals and institutions for providing the intellectual and financial support that made this book possible. But intellectual and financial support were not enough. To embark on this project in the first place required one further ingredient—*chutzpah*. For that, I am indebted to my mother.

Swarthmore, Pennsylvania
August 1985

THE BATTLE FOR
HUMAN NATURE
Science, Morality and
Modern Life

Free to Choose

*If a man does away with his traditional way of living and
throws away his good customs, he had better first make certain
that he has something of value to replace them.*

BASUTO PROVERB

*T*he young women and men who enter col-
lege in the next few years will be students during the two hundredth
anniversary of the Constitution of the United States. They will be
graduating at around the two hundredth anniversary of the adoption
of the Bill of Rights. There is a significant relation between the
aspirations reflected in the Constitution and in most modern college
curricula. The Constitution was designed to affirm and to protect the
rights and freedoms of individuals. It was meant to guarantee a large
measure of individual autonomy in the face of enormous power that
could be imposed by the state. Similarly, a college education—at least
a liberal arts education—is designed to foster and strengthen individual autonomy. It endeavors to create individuals who can think and
act for themselves, often in the face of pressure not only from the
state but from the family, the church, and other social institutions.
Although the Constitution guarantees autonomy as a matter of law,
for most modern Americans, it is only in college that they begin to
experience autonomy as a matter of fact. The liberal arts education
is liberating.

I teach college students, and the impending constitutional bicen-
tennial has gotten me to think about what I would say to orient a

class of incoming freshmen about to begin their education. What would I tell them is the point of it all? What should their educational goals be, and what role will their education play in the rest of their lives?

In attempting to answer these questions, I would focus on freedom. I would tell the students that they live in a time and a place where it is possible for individuals to experience a degree of freedom that their ancestors could not have imagined. Sure, the Constitution has guaranteed certain freedoms for two hundred years. But it is one thing to have freedom by *law* and quite another to have freedom by *more,* by social custom. And while legal freedom has virtually defined the United States from its beginnings, freedom by social custom is only a few generations old.

Americans have always been free by law to decide for themselves what is important and valuable in life. They have been free to choose how to live their lives: what kind of work to do, where to live, who to live with, who to love, what kinds of things to have, and what kind of spiritual or familial commitment to make. But until recently, most of these choices were made by default. People lived where their parents lived, worshiped in their parents' church, worked where their parents worked, married the girl (or boy) next door, and pursued those material things their parents had pursued. Of course, there were differences between generations. By and large, each new generation was better educated than its predecessors and achieved greater material success. What for the parents were often only dreams, for the children became realities. Importantly though, parents' dreams and children's realities were the same. The children were more fortunate, and went further, perhaps, but their parents would have done the same thing in their place.

The making of significant life choices by default has been changing. Probably, the change has been continuous over America's two hundred years, but in the last twenty years or so, the change has been quite dramatic. It is a legacy handed down to this generation of college students by their parents that nowadays, anything is possible; everything is up for grabs. No choices are made by default. The door is open to all kinds of occupations. Liberalized social practices have made it acceptable to have intimate relations with several people of the opposite (or the same) sex before settling on a marital partner. Furthermore, there is no longer any expectation that marriages are forever. They are subject to continual reevaluation, with the possibility always open that they will be terminated if they stop being mutually satisfying. There are no longer clear expectations about relations between family members. Some families stay close; others drift apart. As with marriage, commitment to family life has become a matter of genuine choice. People cannot divorce their parents, but allegiance and respect must be earned; they are not automatic. Finally, people must decide whether to pursue a spiritual life, and if so, what kind of spiritual life. The religion people are born into is not the one they are stuck with. They can convert, be born again, even create their own, personal religion.

I would tell all this to incoming college students. I would tell them that whereas their ancestors were born into powerful, highly structured traditions, or "communities of memory," that largely circumscribed their possibilities, they were not. Indeed, this modern freedom of choice is reflected in the college curriculum itself. A century ago, a college education entailed a rather fixed course of study. One of its principal goals was to educate people into their communities of memory—to make them citizens, with common values and aspirations:

> It was the task of moral philosophy, a required course in the senior year, usually taught by the college president, not only to integrate the various fields of learning, including science and religion, but even more importantly to draw the implications for the living of a good life individually and socially.

This is no longer true. There is no fixed program of study. There is no required course in the senior year. There is no attempt to teach people to live the good life, for who is to say what the good life is? College professors teach what they know and hope that somehow each pupil will turn those lessons into a personal vision of the good life. Furthermore, a significant part of what used to be taught as moral philosophy is now taught under a different name: social science. Social sciences like economics, sociology, and psychology hardly existed as disciplines a century ago. Today, they have a prominent place in every university.

And the change from moral philosophy to social science is more than just a new name for an old subject. Along with the name change has come a change in content. Specifically, the moral component of moral philosophy has been removed. While moral philosophy attempted to teach people how they *should* live, what they *should* value, what roles they *should* play in their communities, social science teaches people how they *do* live, what they *do* value, what roles they *do* play in their communities.

This transformation, from moral philosophy to social science, captures vividly what this book is about. It is about a struggle between the language of science and the language of morality for hegemony in describing what it means to be a person. It is a struggle that the language of morality is losing. The scientific account of what it means to be a person provides much of the conceptual underpinning for our modern freedom of choice. It claims to have much to say about how people do live their lives. But it also claims that neither it, nor any other discipline, has much to say about how people should live their lives. Matters of "should" are very much up to the individual. Thus, the modern citizen experiences a very extensive, and very real, freedom of choice.

The task before us is to examine carefully the scientific account of what it means to be a person: the conflict between the language of science and the language of morality; the consequences of embracing the language of science; and the extent to which the language of science has earned our genuine allegiance. This examination is important because, increasingly, it is the language of science to which new college students must turn for lessons about how to live their lives. It is therefore critical to know what those lessons are, and whether they are the lessons that students should be learning.

Although we live surrounded by scientific-technological wonders, there is good reason to doubt the power of the language of science when it is turned to human affairs. Reason for doubt comes from the fact that the current generation of college students is not the first to enjoy such extensive freedom of choice. The trail they are on was largely blazed by their parents. And the first generation of truly liberated Americans appears not to have done too well with its freedom.

As has been widely reported, many members of the first generation of liberated Americans are experiencing—no, have invented—a collective midlife crisis. Their jobs are not all they had hoped they would be. The material rewards of their jobs do not yield satisfactions that quite match expectations. Their marriages are coming apart. They are not enjoying the close, friendly relations with their children that they thought would be the product of modern, enlightened childrearing. Home is no longer, as it was for Robert Frost, "the place where, when you have to go there, they have to take you in." They feel disconnected from their communities and find themselves with many acquaintances, but few friends. They are distrustful of their political leaders and cynical about the moral principles for which their country stands.

What has gone wrong? How can there be so much dissatisfaction

in the face of so much opportunity? An answer to this question is
suggested in a recent book, *Habits of the Heart,* an analysis of several
hundred detailed interviews with Americans from all over the country, with a wide range of backgrounds, aspirations, and life paths.
The unease that grips modern Americans is clear in these interviews.
It is not so much that they are unhappy as that they are unsure. They
don't seem to know where they belong. They don't seem to know that
they are doing the right things with their lives. They don't seem to
know what the right things are.

The source of this uncertainty lies in the phrase that gave the book
its title—"habits of the heart." In the 1830s, Frenchman Alexis de
Tocqueville published *Democracy in America,* a mixture of observation and analysis of American life at that time derived from his own
recent tour of the country. In it, he observed that what might be
called the first language of America, as embodied in the Constitution
and as lived in daily life, was individualism. America was a collection
of social atoms, each in pursuit of his own interests. In America,
individualism was a matter of principle, a matter of right. But this
individualism was also tempered, by several "second languages" that
united individuals into communities and bound them together. These
second languages—one of religious conviction and one of civic virtue—were sources of moral tradition, of social mores, of habits of the
heart. They were what made it possible for people in pursuit of private, individual interests nevertheless to share public, communal
purposes. The price of modern, liberated America seems to have been
these habits of the heart.

People with a common religious or civic tradition could live in genuine communities, in which at least some convictions, aspirations,
and motives were shared. Their tradition could define for them a
goal, or *telos,* something that told them what the point of life was—
that could be taught in their senior year in college—so that they could
tell whether they were on course, and they could help one another
stay on course. Their tradition could also serve to restrain the pursuit
of self-interest. On this, de Toqueville was clear. He said, "The main
business of religion is to purify, control, and restrain that excessive
and exclusive taste for well-being." It was these second languages of
religious and civic virtue that convinced founders of the Republic
that European philosophers like Montesquieu were wrong to think
that a free society, not run by aristocrats, would run amok. Virtue
could substitute for aristocracy in guiding society.

De Tocqueville was also clear that the first and second languages of America—individualism and religious and civic virtue—would be in constant tension. There was no guarantee that the traditions that were keeping America civilized when de Tocqueville made his tour would remain strong without eternal vigilance. And Americans have not remained vigilant. In choosing to be free to choose, we have let traditions weaken. In consequence, modern society *has* run amok. In an attempt to restore order and purpose to modern life, our current political leaders appeal to us to display those virtues that are part of our national heritage. They appeal for a return to the "old values" of religious commitment and civic participation. And their appeals are not falling on deaf ears. Many people are struggling to reestablish a communal, public life. America is undergoing a religious revival. Its colleges are beginning to experiment again with required programs of study.

But as *Habits of the Heart* documents, these efforts, for many, are marked by a deep, underlying confusion. It is not that people no longer possess their traditional second languages. Instead, the problem is that the second languages are slowly being infused with terms from the first language. Civic participation is increasingly understood to be nothing but the pursuit of self-interest in the public arena. Only rarely is one expected to submerge his interests for the common good. So explicit has this understanding of political participation become that the "lobbyist" has moved from the smoke-filled back room onto center stage. The lobby is now an official, political institution. Similarly, for many, religion is increasingly understood to be an individual, private affair, whose function is to serve the idiosyncratic needs and desires of those who participate in it. Where once most people belonged to "the church," a public institution that always had its eye turned to the secular world, with the aim of participating in it and influencing it, now many regard the church as unsatisfactory and turn instead to sects, which set themselves apart from the world, or even to individualistic mysticism. Americans are no less "religious" than they ever were—95 percent of Americans say they believe in God. It's just that being religious no longer means what it did in de Tocqueville's America.

The infusion of our second languages by the language of individualism affects people's lives in many different ways. It can explain much of the disappointment people experience with work. It is possible to distinguish three different kinds of work: work as a job; work as a career; and work as a calling. Those who have "jobs" don't expect much satisfaction from them. They do them for the material benefits they bring, and nothing more. Progress, advancement, and success are all measured by wages. Those with "careers" are more fortunate. While people with jobs are just working, people with careers are going somewhere. There is a clear trajectory of advancement that defines success, and people derive satisfaction from advancing over and above the material rewards that it brings. But the trouble with careers is that not everyone can keep advancing. Some people reach the age of about forty, and the realization hits that they are stuck. They won't be getting to the top, as they had imagined in their youthful, ambitious fantasies. They face twenty-five more years of just holding on, with nowhere to go but down. Small wonder that they experience a midlife crisis. Those with a "calling" are the most fortunate. For them, it is the concrete products of their work, and not just personal advancement, that provide satisfaction. People with a calling are doing something of value. It will not lose its value even if they are stuck doing it, with no prospects for moving up, for twenty-five or more years.

The trouble is that for work to be a calling—to be valuable—there needs to be a communal understanding of value. Teaching the same material year after year, treating the same illnesses, writing the same wills, farming the same crops—whether these activities are jobs or callings depends upon whether their products are generally regarded as valuable. And communal understandings of value derive from the second languages of America, not the first. In the language of individualism, what is valuable is very much an individual affair. Thus, as the second languages of America give way to its first language, callings give way to careers and to jobs.

The disintegration of second languages also affects the character of close personal relations. In the *Nichomachean Ethics,* Aristotle discussed the essential ingredients of friendship. Two of them are obvious. Real friends enjoy each other's company and provide for each other's needs. The third ingredient is less obvious but, according to Aristotle, most important. Real friends share a common moral or ethical commitment. Thus, a friend helps keep one on the right path. A friend is critical even while providing unconditional affection. Real friendship is the very opposite of flattery. Friendship, understood in this way, is the glue of a good and cohesive society. Friends are not to be confused with tennis partners and dinner party companions. But friendship is undercut by the language of individualism, since individualism undercuts the very notion of common moral or ethical commitments. As pointed out in *Habits of the Heart,* the closest that many people come to having real friends is in the relations they have with their therapists. Therapists are "friends for hire." They will provide honesty instead of flattery, if that's what their clients want. But of course, while therapists are their clients' "friends," their clients aren't theirs. The therapeutic relation, unlike genuine friendship, is both asymetric and circumscribed. But at the moment, it's the best that many people can do.

Furthermore, the therapist is only a very limited friend in another important way. The therapist can help people *feel* good, but she can't help them *be* good. The therapist can help people do *well*, but she can't help them do *good*. That is because the therapist shares in the language of individualism, according to which what it means to *be* good and to *do* good is up to each person as an individual.

So it is, I would tell the incoming students, that their parents have not fared well with their unprecedented freedom of choice. They have had what the first governor of Massachusetts, John Winthrop, called "natural freedom," the freedom to do whatever they wanted. But in gaining this natural freedom, they have given up what he called "moral freedom," the freedom to do what is good. For the price of natural freedom is that no one is sure what "good" is anymore. This seems like too steep a price to have to pay.

Can today's college students somehow reverse the erosion of America's second languages? Can they reinvigorate public morality, the traditions of religious commitment and civic virtue? Not, as we've seen, with what they are about to learn in college. For like work, friendship, the family, and the church, college has fallen victim to the language of individualism. College professors have no clearer idea of what the good life is than the rest of society. They offer the language of science rather than the language of morality. The scientific account of what it means to be a person provides the conceptual underpinning for the modern commitment to individualism. If the grip that this language has on our lives is to be contained, we must be able to understand and criticize the conceptual foundations that give it strength. Perhaps in this way, we can start turning natural freedom back into moral freedom. We shall see.

ple's agendas these days. They are the subject of intense debate and
disagreement, of controversial legislation, of lawsuits, of potential
constitutional amendments, and of Sunday sermons. They divide
nations, communities, and families. And while stated in specific form
they seem utterly modern; stated more generally, they have concerned people for centuries. How should society be organized? How
should the resources of society be distributed among its members?
How much should individual freedom be restricted, and in what ways?
What is the extent of our responsibility to other human beings, and
to the society to which we belong? What is the proper mode of human
conduct, and how should it be instilled in people? If we were clear
about the answers to these general questions, some of the more specific ones of current concern might never have arisen; or if they arose,
we might collectively puzzle out the answers to them without getting
the feeling that we were speaking different languages. But we don't
seem to know the answers to the general questions. Or perhaps more
accurately, lots of people know the answers to the general questions;
it's just that they all know different ones.

Although centuries of concern about these questions have failed to
produce consensus about the answers, people do agree that these are
all reasonable questions to be asking. The questions all make sense,
and they are significant enough to justify the anguish, frustration,
sweat, and even blood that have been spilled in trying to answer them.
What the questions have in common is that they are all *moral* questions. They all involve figuring out how people, families, schools,
communities, and states *should* act. They are all concerned with spelling
out right and wrong, good and bad. And people seem to agree, as they
have agreed for millennia, that it is appropriate to ask moral questions
about human beings and human social institutions.

This all must seem obvious, but it is important to realize that from
the perspective of our modern understanding of the world, a truly
distinctive thing about human beings is that it is they, and only they,
about whom it makes sense to ask moral questions. It doesn't make
sense to ask whether planets *ought* to revolve around the sun, whether
plants *ought* to grow green leaves, whether bees *ought* to sting intruders at the hive. Planets simply *do* revolve around the sun, plants *do*
grow leaves, and bees *do* sting. That's all there is to it. But in the
case of people, knowing what they *do* isn't enough; we also want to
know whether they *ought* to be doing it. We carve the world into two
domains. One is the domain of facts, of knowledge of the way things
are. The other is the domain of morals, of knowledge of the way

things ought to be. As a result of the work of philosopher David Hume, in the eighteenth century, we have been exhorted not to confuse these domains, to keep them distinct. About facts we can ask, true or false; about morals we can ask, good or bad. The domain of facts properly includes all aspects of the world around us. We can, and do, gather facts about the motion of heavenly bodies, about the growth of plants, and about the behavior patterns of animals, including people. We even gather facts about morals, about what people think is good or bad. But facts take us only so far when we are focused on ourselves. For in addition to knowing what people *do* think is good or bad, we want to know what they *should* think is good or bad. In addition to knowing that the majority of society is against capital punishment, we want to know whether they *should* be against capital punishment. It is this set of concerns about "shoulds," about moral issues, that sets the questions we ask about ourselves apart from the questions we ask about the rest of the universe.

The Relevance of What Is to What Ought to Be

When Hume was instructing us about the distinction between what is and what ought to be, he warned us against making a particular type of mistake when thinking about moral questions. The mistake has come to be called the "naturalistic fallacy." In substance, the mistake is to move from factual premises to a moral conclusion, to infer what ought to be from what is. Nothing that is known about how people actually, in fact, *are* justifies drawing any particular conclusions about how they *ought* to be. Knowing, for example, that people have a natural tendency to show care and concern for others doesn't help in deciding whether they should show such concern, or how much of it they should show. Neither would knowing that people have a natural tendency to be selfish. Knowing that a majority of people oppose capital punishment doesn't help in deciding whether they should oppose it. Decisions about how people should act simply have to be made on other grounds. Then, knowing about how people actually are may suggest how likely they are to behave as they should: how likely they are to be moral. Knowing the facts of the matter may help to determine just how hopeful to be that a moral society is possible, but they won't tell what it would mean for a society to be moral.

In everyday discourse, the lesson of Hume's dictum about the independence of "ought" from "is" is not unimportant. For example, imagine a breakfast table conversation between a father and a young son who has just found out that his baseball player idol doesn't really eat the breakfast cereal that he hawks incessantly on television. "Listen," the father says, "the company pays him a lot of money to say those things. It's hard for him to pass up the opportunity to make money while he can. In a few years, he'll be out of baseball, and no one will even remember who he is. It's only human nature to take advantage of opportunities to make money when you get them. Besides, most people don't really believe it when athletes and movie stars say they eat this cereal, or wear those sneakers." All true, perhaps, and this appeal to "human nature" might provide an explanation of the athlete's false endorsement. But it won't, as the son will know, and it can't, as Hume argues, provide a logically sound justification. The mere fact that it is human nature to sacrifice honesty for profit under some circumstances doesn't make it right.

But even if understanding what is doesn't carry any strict implications about what ought to be, when people go about the business actually of deciding what ought to be, certainly they are influenced by the way they think things are. Calling some things good and some things bad depends upon having a notion of what the good life for people looks like, on having some ideas about human possibilities under the best of circumstances. And a notion about what the good life is in turn depends upon a notion of what people are. Thus, the "facts of the matter" have an important role to play in resolving moral uncertainties. Although they may not dictate oughts, they surely constrain them.

Imagine, for example, trying to formulate a code of moral conduct for a colony of ants. In principle, the same kinds of issues could be raised about ant morality as about human morality. One could worry both about the individual ant's rights and responsibilities to the colony and the colony's rights and responsibilities to each individual ant. That such worries don't arise stems from what people know (or think they know) about what it means to be an ant, in contrast to what it means to be a person. But a visitor from outer space who watched colonies of humans and ants, each living together, working together, dividing responsibilities, sharing resources, caring for offspring, and so on, might have a hard time figuring out which colony lived by a moral code and which did not. So a good deal of how people think about moral questions—indeed, a good deal of how they decide what is or is not a moral question—depends on what they know, upon the facts.

We can see how various kinds of knowledge about what it means to be a person might help resolve current moral controversies. If people *knew* when the fetus became a person, it might help distinguish moral from immoral abortions. If they *knew* when children attained the level of rational competence that allowed them to make decisions for themselves, they might be able to decide at what point parents no longer had the right to control their children's lives, and schools no longer had the right to dictate what children learned. If they *knew* how to tell whether a person had or had not lost the capacity to think clearly, they might be able to decide at what point a person's right to refuse life-sustaining medical care should be abridged.

In none of these cases is the moral question simply and certainly solved by this knowledge. A society might decide that people never (or always) have the right to kill themselves, independent of their mental competence. And in this respect, Hume was right that "ought" doesn't follow automatically from "is." But in each of these cases, possessing the relevant knowledge would help people see their way more clearly to a resolution to the moral debate. If nothing else, such knowledge would rule certain kinds of moral arguments out. Whatever one thought of abortion, it couldn't be called "murder" if it were known, by an agreed upon definition of life, that the fetus was not alive until, say, the last trimester of pregnancy. So what we think people ought to be and do depends in part on what we think we know they are and can do. Resolution of moral debate depends on people's conceptions of human nature, on what they think it means to be a person. Where, then, do these conceptions of human nature come from?

The answer is that they come from lots of different places. They come from our individual, everyday experience, from the so-called school of hard knocks. As people interact, they discover regularities in the way others react to certain situations. From these discoveries, they build up a picture of what a typical person wants, does, and expects them to do. This is so commonplace, and so inevitable, that it is hardly noticed. When a child first starts going to school, he may expect that teachers will be like his parents—hanging on his every word, consumed with his welfare, eager to make him happy. He quickly learns that teachers aren't that way, that being a parent is something special. When an adult enters the world of commerce, she learns to be suspicious of people who seem to be offering something for nothing. She learns that the point of commercial transactions is to get something you want, and she starts shrewdly looking for "what's in it for him" when she is offered a deal that seems too good to be true.

The picture of human nature people build up with experience serves as a crucially important guide as they encounter new people in new situations. How much they depend upon it really hits home when the picture is violated. When people are forced to interact with someone who suffers from severe psychological disturbance, they don't know what to do. All expectations about what people want and how they think and act are violated. They stand helplessly, not knowing what casual remark or act will induce violent rage or hopeless despair.

While personal experience is helpful, it has its limits, and if it were all people had to go on in constructing a conception of human nature, they'd be in trouble. It takes time to accumulate experiences, it is not always easy to interpret them, and everyone has different ones. If someone had to learn about the profit motive the "hard way," he'd lose his shirt before he learned. Experience is buttressed by a set of social institutions into which people are born, institutions whose very shape is itself guided by an operative conception of human nature. There is first the family. By socializing children, by teaching them how to behave, parents are teaching what is expected of people and, by implication, what it means to be a person. In fact, children may learn especially important lessons about human nature from what their parents teach them *not* to do. From their parents' insistence that they shouldn't lie, or cheat, or steal, or fight, children may draw the conclusion that people, by their very natures, are tempted to do all these things. After all, children don't hear their parents admonishing them not to cut off and eat their toes, or not to throw their best friend's baby brother out the window. There are all kinds of things that people shouldn't do, but parents focus their attention on the particular "shouldn'ts" that they might do.

The same could be said of the other social institutions from which people derive their picture of human nature. The church, the school, the state all contribute to socialization both by teaching what people should and shouldn't do and by teaching what they might do. In general, the dictates of our social institutions teach both about the noblest of human aspirations and about the basest of human frailties. And by absorbing these dictates we absorb a conception of human nature which guides us in filtering and interpreting our own particular everyday experiences with other people. To attempt to locate oneself in the world without these guiding institutions, to attempt to build up, simply through experience, a set of expectations about other people and a set of guides to our personal conduct is a hopeless task. It is to be set adrift, to experience what novelist Milan Kundera describes as "the unbearable lightness of being," to float, without anchor or direction, through space and time. Few people have to experience this dislocation. Most of us inherit, inevitably, the prescriptions and the proscriptions of our social institutions. With these, the development of knowledge about what human nature is becomes a more manageable task, and from that knowledge, a conception of what people should be takes shape.

What I am suggesting, then, is that the way concerns about how people should act are resolved is significantly affected by what socializing institutions and individual experience tell us people are. I don't mean to suggest by this that everyone walks around with a conception of human nature on the tip of the tongue. Very few people have ever even thought about "human nature," let alone developed a coherent and articulate picture of it. Nevertheless, people are not paralyzed when faced with a moral decision, nor for that matter are they helpless and confused when faced with figuring out how someone is going to act in a particular situation.

What I have in mind in suggesting that our conception of human nature informs our moral decisions is that most people have an *implicit* conception of human nature. Experience with people and with social institutions like the family, the church, the school, and the state engenders a set of expectations, or rules-of-thumb about how people act, and about how they are capable of acting. Among the functions that this conception serves is to indicate which sorts of human action are "normal," that is, in keeping with human nature, and which ones are surprising or unusual. One hears a story about a woman who held three jobs, never spent a penny on herself, and managed to put four children through college. It's a nice story, but not especially surprising. Everyone knows that parents make all kinds of sacrifices for their children. It's only human nature. Hearing essentially the same story, except that the woman gave her money away to strangers, or spent it on her cats, produces surprise and puzzlement. The woman must be crazy. Some things give us pause and get us to look for more information. Other things are run of the mill, routine, unsurprising. Which is which is to a large extent influenced by what we think (implicitly) people are.

Aspects of this implicit theory of human nature that are usually deeply hidden come to the surface when evolving social practices require people to change them. For example, if asked, twenty years ago, to enumerate the motives of professional athletes, loyalty to the team would have been high on the list. Dedicated service, game after game, and year after year, ignoring pain and injury, was what was expected. Those occasions in which these expectations were violated, as when an athlete "held out," or refused to play, until he got a sweeter contract, commanded attention—and disapproval. How could an athlete disregard the best interests of the team just so that he could line his own pockets? Why it was, it was . . . unprofessional. Many an otherwise revered athlete suffered severe public censure for being unprofessional in this way. Nowadays, of course, all this has changed. The common expectation now is that a professional athlete is loyal only to himself and that he will eagerly sell his services to the highest bidder. "Professional" has come to mean "in it for the money." Nowadays, what demands an explanation is the rare occasion in which an athlete refuses to sell out to the highest bidder. The rule-of-thumb has been altered, from one that featured loyalty to one that features greed. What used to command disapprobation and demand explanation is now taken for granted, and what used to be taken for granted now demands explanation.

On a much larger scale, a similar change in operative rules-of-thumb has evolved regarding differences between men and women. In modern society, women who are committed to careers and men who do their share around the house are hardly noticed, while women who elect to be housewives and men who rule the house with an iron hand require explanations. A generation ago, of course, the reverse was true. And somewhat more slowly, ideas about patterns of social and sexual interaction are changing, too. Women who take the social or sexual initiative, asking men out and seeking satisfying sexual relations, are becoming less and less a phenomenon in need of explanation.

These changing ideas about the differences between men and women illustrate how decisions about human morals are in part the product of what people take the facts about human nature to be. Many of the moral issues that have been debated over the years—issues concerning the rights and responsibilities of citizens as against the rights and responsibilities of the state—have historically been answered in one way for men and in a different way for women. At different points in history, women have been denied the right to vote, the right to own property, the right to an education, even the right to some control over what happened in the marriage bed. Within the framework of a theory of human nature that treated certain differences between men and women as fundamental, none of these practices would have been regarded as immoral. Indeed, they would hardly have been noticed as practices requiring an explanation. That a host of individual rights for which American colonies had fought a revolution were routinely denied to a major segment of the population would simply have been viewed as in the nature of things—in human nature, no more worthy of comment than the fact that children weren't accorded the same rights as adults.

The fact that rules-of-thumb regarding differences between the sexes change does not mean that all differences in practice immediately are eliminated. But what it does mean is that the burden of proof has shifted. Now, differences in rights and responsibilities between the sexes are the subject of debate rather than presumption. Now, differences in treatment must be defended. Now, the possibility is open that society's treatment of women has been immoral. When our understanding is that women (and children, and blacks) are, by their very natures, a different kind of (perhaps more limited) moral agent than men, our moral demands on them and expectations of them will reflect that difference. When our understanding grants women (and blacks) equal moral status, our judgment of what constitutes right and wrong treatment of them must change accordingly. So it is in general that our understanding of human nature affects our moral decisions.

Most people agree that all citizens are entitled to the individual freedoms guaranteed by the Bill of Rights. People are entitled to speak freely, to read what books they want, to assemble freely, to bear arms, and so on. Yet, each and every one of these freedoms is abridged in the case of children. Why? Clearly, it is because people don't think children possess whatever knowledge, rationality, responsibility, or morality is required to be accorded the full rights of citizenship. Even in the context of complete agreement about what rights citizens are entitled to expect, our conception of human nature influences us to deny those rights to children. Furthermore, it will influence how children are reared in the home, how they are educated in the school, and when in the course of their development they are welcomed into the adult community. It can always be asked, of course, whether our children are being treated morally, but the answer we get will not be based strictly on moral argument. It will also depend on just what sorts of creatures we think children are. The treatment of children, like the treatment of women, is a case in which what people think *ought* to be is affected by what they think *is*.

Science

We have seen that rules-of-thumb about human nature are built out of everyday experience together with the guidance provided by social institutions like the family, the church, and the school. Indeed, traditionally, these have been the principal sources of knowledge about the world in general. But over the last three hundred years or so, tradition has given way to a new authority: the authority of science. People no longer seek their priests or grandmothers to find out how to treat a rash or a stomachache. Nor do they turn to them to find out why the sky is blue, or how cows turn grass into milk. It is firmly understood by virtually all citizens of industrial societies that the ultimate authority in certain domains is neither the priest, nor the parent, but the scientist.

As science has developed over the last three centuries, the domains over which it has sought authority have expanded. The initial achievements of modern science were in its ability to explain the behavior of nonliving things, like the planets and other heavenly bodies. This was the concern of the physics ushered in by Galileo and Newton in the seventeenth century. But even as people were willing to defer to the authority of the scientist with respect to the physical world, it was believed that physics went only so far, that living things shared some special characteristics that inanimate things did not possess and that physical principles could not explain. Thus, the authority of science stopped at life. With the emergence of scientific biology in the eighteenth century, however, the explanatory domain of science was extended. At least some characteristics of all living things could be understood scientifically. Indeed, perhaps all characteristics of nonhuman living things were susceptible to scientific analysis. But still, people were thought to be special. The human capacity for reason and the human soul placed human beings outside the bounds of physics or biology. In the last one hundred fifty years or so, this last barrier to scientific authority has come under steady attack. The "human sciences"—economics, sociology, anthropology, and psychology—are attempting to show that no aspect of human life can resist the power of scientific scrutiny. As these disciplines progress in explaining people in the same way that physics has succeeded in explaining planets, they come to replace the family and the church as the sources of our conception of human nature. Our everyday conception of human nature comes to approximate the scientific one.

To examine what a "scientific" conception of human nature is, we

must first look at what characterizes scientific conceptions more generally. For most people, the first encounter with "science" is as a
subject in school. Somewhere around the fourth grade, a portion of
the day gets set aside for science lessons. These may be lessons in
nature—folksy, Disneyesque accounts of exotic creatures living in
exotic lands, or perhaps nowadays, discussions of food chains, pollution, endangered species, and ecology. Or they may be lessons in
"chemistry"—discussions of atoms and molecules and of what happens when different elements are combined or when substances are
heated. Or perhaps they may be lessons in "physics"—in how flash-
lights work, or how airplanes stay in the air, or how ships manage
not to sink. The impression that this regimen of science lessons
engenders is that "science" is a body of knowledge, that what sets it
apart from "history" or "English" is what it is about. As schooling
continues, people learn that just as there are different kinds of history (world history and American history) and different kinds of English
(grammar, vocabulary, poetry, novels), there are different kinds of
science. The scientific monolith of early schooling is differentiated
into domains like biology, chemistry, and physics. Still, however, science is understood to refer to some domain of information.

   Although most people never move beyond this understanding, it is
fundamentally mistaken. What is distinctive about science is not so
much the things that science is about as it is a set of procedures, or
methods, for finding out about those things and a set of standards, or
criteria, for deciding whether anything has actually been found out.
In other words, "science" is really a process for finding things out
about the world. It is a process that, *in principle,* could be applied to
any set of phenomena. There could be a science of cooking, or pole
vaulting, or gardening, just as there are sciences of biology, chemistry, and physics.

   One of the most distinctive qualities of human beings is their persistent effort to make sense of the world around them. Whether children or adults, farmers or professors, members of highly technological
or of primitive cultures, a substantial part of people's daily activity
involves attempting to understand the events that affect their lives.
Science is just one particular approach to seeking an understanding
of the world. The reason it commands special attention is that it has
been so powerful in enabling people to intervene effectively in their
environments. Every aspect of life in modern, highly technological
society is testimony to the explanatory power of scientific modes of
understanding. Every time someone flips on a light, drives a car, takes

a photograph, turns on the television, photocopies a letter, or computes his income taxes with the aid of a calculator, he is presenting himself with evidence that science has delivered the goods. The kind of understanding that characterizes science is what has led to all of these technological innovations. To the extent that technology enables people to control their environments better than they could before, it validates the scientific conception of the world. But what is the scientific conception of the world? What does it mean to understand something scientifically?

In attempting to understand a given phenomenon, the scientist searches for some other event or phenomenon that caused it. To understand what causal explanations typically involve, consider an example: Suppose someone drops an expensive, antique vase, and it falls to the floor and breaks into a hundred pieces. What caused the vase to fall and shatter?

A first response to this question might be, "He dropped it. That's why it fell and shattered." This "causal" account has several important properties. First, the cause identified was such that, under the circumstances, had the cause not occurred, the phenomenon would not have occurred. Thus, if the vase had not been dropped, it would not have fallen and broken. In short, dropping the vase was *necessary* for it to fall and break. But this cause was only necessary *under the given circumstances*. Had the circumstances been different (for example, an earthquake), the vase might have fallen without anyone's intervention. Second, under the circumstances, the cause identified was *all* that was required to bring about the phenomenon. Thus, once the vase dropped, nothing else had to happen for it to fall and break. Dropping the vase was *sufficient* for it to fall and break. But of course, this cause was only sufficient under the circumstances. If the vase had been dropped while it was over a mattress, it would have fallen, but it wouldn't have broken. Finding causes of a phenomenon, then, involves finding influences that are necessary and sufficient for its occurrence, under a given set of circumstances.

Does this mean that "He dropped it" counts as an appropriately scientific explanation of the behavior of the vase? Not quite. For what is perhaps the most significant feature of scientific explanations is that they relate the specific phenomenon to be explained to other, similar phenomena. At the heart of scientific explanation is the search for *generalizations,* or *laws.* While the fact of the matter is that this particular vase may have fallen and broken after a particular person dropped it, it would have fallen and broken no matter who dropped it. Indeed, the fact that it was dropped is not essential to its falling and breaking. No matter what is responsible for setting a vase in free fall, it will move toward the center of the earth. Furthermore, the object needn't be a vase. Any object in free fall will move toward the earth, and any object composed of certain sorts of materials will break when it contacts the hard ground. Finally, the cause of the behavior of the vase is the same as the cause of the motions of the planets. For when it is said that the vase fell and broke because someone dropped it, what lies beneath this explanation is a wealth of knowledge from physics about gravity and its effects on objects in free fall. It is this knowledge of physics that tells us that the vase would not have fallen and broken if it had been dropped on the moon. It is this knowledge of physics that tells us to what general class of phenomena the behavior of the vase belongs. Without knowing this, someone might still be able to say that the vase fell and broke because it was dropped. But he would not necessarily be able to use this knowledge to help understand other phenomena, or perhaps predict or control events in the future.

How we generalize a particular phenomenon has a lot to do with what we do about it in the future. Science is an attempt to discover which kinds of generalizations will provide the broadest and most powerful guidance about what to do in the future. In suggesting that science has delivered the goods, what I meant was that the generalizations it has provided have enabled people to predict and control the world in which they live with unparalleled success. That is why the kind of understanding reflected by science's search for general, causal laws has become a model of what understanding should be in the technologically advanced world.

That scientific understanding involves generalizations, or laws, is important for another reason. It is only because scientific explanations of particular events are really explanations of classes of events that people are able to evaluate their causal judgments. Consider again the fallen vase. We believe the vase fell because someone dropped it, and gravity took over. Someone else believes the vase fell because a demon in the center of the earth wanted it to. Still a third person believes the vase fell because its owner had done something to anger the gods. Which explanation is right, and how does one know? In fact, any of these explanations might suffice to account for the fact that this *particular* vase fell to this *particular* floor at this *particular* time. And so might dozens of other explanations.

Or consider another example. Suppose a baker follows a recipe to make a cake and the cake comes out heavy and dry. Is it because he didn't beat the eggs enough, or because he baked it too long, or because the oven was hotter than the thermostat setting indicated, or because the recipe was bad? There is no way of telling in the case of this one cake. It could be for any or all of these reasons that the cake was a flop.

The way to sift through this plethora of possible explanations is to make more observations—to gather more data. In the case of the fallen vase, someone might walk around with a careful eye open to note other vases falling, or other objects falling. In the case of the cake, the baker might take note of how other things made in the same oven turn out, or how other people following the same recipe do. Repeated, systematic observation of a phenomenon helps to weed out the implausible candidates for causes and focus on the likely ones. If other cakes baked in the same oven come out fine, it is unlikely that the failed cake was the oven's fault. If vases dropped by other people also fall and break, it is unlikely that this vase fell and broke because its owner angered the gods.

Systematic, repeated observation allows one to begin to separate the wheat from the chaff as far as causal accounts are concerned, but it has its drawbacks. It may be necessary to wait and watch for a long time before collecting enough examples of broken vases to be confident about what caused them to fall and break. This takes time and patience. What if one has neither? Science has provided an alternative. Instead of waiting for vases to fall, we can make them fall ourselves. We can try to *create* additional observations that relate to the phenomenon under investigation. This intentional creation of observations is what scientists refer to as *experiments*. An experimenter might drop other vases, or other objects, and have other people drop them, and she could do this repeatedly, at different times and places. If she did, she would discover that no matter what the time or place, no matter what the object, and no matter who dropped it, it fell. Or the baker might try other recipes in the same oven, the same recipe in different ovens, and so on to determine whether there were any circumstances in which the cake came out all right. In this way, by appropriate observations via experimentation, one could move rapidly and surely to an accurate causal analysis. With such an analysis in hand, it would be possible to produce good (or bad) cakes on demand, or make objects fall, or keep them from falling.

There is no question that systematic observation and experimentation are what have given science its extraordinary explanatory power.

But it is important to realize that the logic of these methods rests on a very significant assumption. The assumption is that other falling objects belong to the same class as the vase and that future cakes belong to the same class as this first failure. Without this assumption, there would be no reason to treat the phenomena observed or created by experiment as relevant to the ones the experimenter initially set out to explain.

This is no small assumption. Every phenomenon or event is in some respects unique. It occurs at a particular place and a particular moment, and once that moment has passed, it can't be recaptured. Even if other cakes come out dry, the baker can't be sure that what makes them fail is what made the first one fail. Maybe there was something special about the butter and eggs used in the first cake. It can never be proven that there wasn't. By asserting that the cake failed because the recipe was bad, the baker is saying that for all the uniqueness of that first cake, made with those ingredients at that time, it was just like any other cake would be that was prepared according to the same recipe. Said another way, when science formulates the sorts of generalizations that it is after and that justify doing experiments, it is engaged in a process of abstraction. It looks at the unique phenomenon, place, and time and abstracts from it those features or elements that it has in common with other phenomena, places, and times.

What this discussion reveals about scientific understanding is that in pursuing causal laws, science tells us in what categories individual phenomena belong. In doing this, science ignores the many features that make each and every object or event unique and focuses on essential properties they have in common. It is just this feature of scientific understanding that allows for the prediction and control of aspects of the future on the basis of what is understood about aspects of the past.

Now the search for causal laws presupposes that the phenomena of nature are sufficiently orderly and repeatable that laws are there to be found. Without this belief in the orderliness of nature, scientific activity would not make much sense. When talking about falling vases, this belief seems quite straightforward and uncontroversial. That their uniqueness as objects that fall can be disregarded is hardly a surprise. But as we move into other domains, uncertainty arises. Consider, for example, the claim that people cheat on their income taxes because it is human nature to pursue self-interest, without regard for the welfare of others or for one's responsibilities to the larger community. An account like this says that for purposes of causal analysis of tax cheating, the uniqueness of individuals can be disregarded. All people can be treated as members of a single class that has in common the pursuit of self-interest. Clearly, in a domain like this, treating unique individuals simply as members of larger classes is not so straightforward.

Indeed, even treating human actions as phenomena, like the motions of vases, that *have* causes is not so straightforward. Most people hold the view that the appropriateness of scientific understanding stops at phenomena that are essentially human. In their everyday lives people do not seek to understand their own behavior and that of their friends, loved ones, employers, and political leaders in terms of necessary and sufficient causal laws. They think of others and themselves as having control over their actions, as exercising discretion and choice, and not as being pushed around, like vases, by necessary and sufficient causes. Are human actions caused? Are human actions reliable and repeatable in the way that the action of the falling vase is?

In general, when the attempt is made to understand human action rather than the behavior of inanimate objects, notions of purpose, belief, expectation, and deliberation replace notions of necessary and sufficient causes. People are comfortable with explanations of human action in these terms and uncomfortable with explanations that have a scientific character. People assume that there are *reasons* for what they do, and the task in understanding is to discover those reasons. And because how we understand something has practical consequences for what we do about it, the commitment to explanations of human action based on reasons is reflected in many customary social practices and cultural institutions. For example, it is common to try to influence people by *reasoning* with them, by giving them reasons to change their minds. It is common to hold people responsible for what they do. They are praised or blamed when appropriate, their actions are regarded as moral or immoral, they are sent to prison, or elected to high office. But no one would consider holding the vase responsible for falling. This reflects a commitment to the belief that vases can be understood in terms of causal forces, but that people cannot be understood in the same way. Thus in our everyday understanding of human action—of human nature—we are usually content to bypass the explanatory power offered by the sciences.

A *Science of Human Nature*

Despite our reluctance to replace our everyday explanations of human action with scientific ones, the power of science has been engaged in a relentless assault on the human barricades. If physics can give us the power to fly, if chemistry can give us food preservatives and life-saving drugs, if biology can give us the means to detect and prevent birth defects, who knows what a science of human nature can deliver? Our everyday conception of human nature may soon be regarded as nothing more than a superstition, which the methods of science will allow us to peel away.

What might a science of human nature look like? Remember that the goal of science is to discover causal laws, generalizations that apply to a wide range of different phenomena at different times and places. The goal of a science of human nature would be to discover generalizations that were true of all people. At first glance, the task of discovering universal generalizations about human nature seems positively daunting. Even the most sheltered observer of human nature must be struck by the wealth of human diversity. People living in cities behave differently from people living in the country. Europeans are different from Asians. Adults are different from children. Men are different from women. It is hard to believe that science could cut through all this difference to find underlying similarity. It is tempting to conclude that there is no human nature, that there is no set of characteristics that all people, in all places, at all times share. What people are, and how they act, will differ enormously from time to time and culture to culture. Human behavior is dramatically affected by the particular circumstances in which people find themselves. The physicist knows that a falling rock is a falling rock is a falling rock. The biologist knows that a rose is a rose is a rose. But it just isn't so that a person is a person is a person. Pursuing general laws of human nature is sheer folly. Or is it?

It isn't really true that all falling rocks are the same. In the ideal world of a vacuum, with no air or wind resistance, every rock, no matter what its size or shape, will fall at the same rate. So will leaves, feathers and pianos. But in real life, size and shape make a difference. What the physicist does in arriving at a generalization about falling rocks is treat them as abstract objects falling in an abstract universe. Factors like wind are regarded as a nuisance, giving the impression that each object falls in its own unique way and obscuring what all falling objects have in common. While it is true that the wind and the air resistance will combine with the size and shape of every particular object to give it its own distinctive motion as it falls to the ground, these idiosyncrasies are only part of what determines the way an object will fall. The rest is determined by gravity, which affects all falling objects in the same way.

So a complete understanding of the falling of rocks has two parts. There is first the general part, the laws of physics that apply indiscriminately to all falling objects. And then there is the specific part, which takes account of the interaction of each unique object and the environment in which it is falling. Perhaps the same story can be told about people. The explanation of human action may also require two parts: one part that consists of the universal characteristics of human beings; and a second part that accounts for the way in which these universal characteristics are affected by the influence of particular times and places.

When a physicist wants to get at the general laws that are involved when an object falls, he eliminates the specifics. He eliminates the effects of air and wind by creating a vacuum and studying how objects fall in the absence of extraneous influences. By analogy, to get at human nature, one would want to study people in a "vacuum," an environment free of the various particular influences that give people their uniqueness. While the physicist's vacuum eliminates air and wind, the human vacuum would have to eliminate culture. So the question would become, what are people like in the absence of culture? How does a person act in the state of nature? What is "natural man"?

These questions cannot be answered by doing experiments. We cannot take babies, put them by themselves in the woods, and see what they come out like as adults. That is, cultural vacuums can't actually be created. However, this practical limitation does not leave the scientist without resources. He can observe and experiment with other animals in the state of nature. He can see what they are like and extend his findings, by analogy, to humans. And he can speculate. He can construct hypotheses about what sort of human nature *would* be revealed if people were forced to fend for themselves, without the support and influence of a social environment. The logic underlying this attempt to uncover what natural man would be like is the same logic that underlies the physicist's experimentation. Neutralize or eliminate the forces that make for individual uniqueness, and what is left is universal. What is left—what is natural—is human nature.

Among those who have asked what people would be like in the state of nature—what human nature divorced from the influence of culture is—the person whose ideas have had perhaps the greatest influence on the modern conceptions that will concern us in this book is Thomas Hobbes. Hobbes lived in the England of the seventeenth century, at a time when modern science was just taking root. In *Leviathan,* he offered a theory of human nature that was at the same time a defense of a particular kind of political organization. Hobbes thought that each human machine "endeavors to secure himself against the evil he fears, and procure the good he desires." Moreover, the human machine never rests in this pursuit. "Felicity of this life consists not in the repose of a mind satisfied. . . . Nor can a man any more live, whose desires are at an end. . . . Felicity is a continual progress of desire."

In other words, people are moved to action so as to satisfy their desires, and their desires can never be fully satisfied. They will always want something. The reason for this unlimited desire, Hobbes asserted, is that while it is fixed and universal human nature to try to satisfy desires, what people actually desire is neither fixed nor universal. What people want is conditioned on what others around them have. Value and worth "consist in comparison. For if all things were equally in all men, nothing would be prized." Thus, to satisfy desire is to have more than the people around you have. But because satisfaction of desire will mean this to everyone, there will be endless competition among people to outdo one another. And the competition will be ugly. For the best way for one person to outdo another is simply to take what is his. And the best way to do that is by being powerful. Thus, there is in men "a perpetual and restless desire for power that ceases only in death." It must be perpetual and restless, requiring eternal vigilance, for if one person lets down his guard, even for an instant, someone else will be there to take what is his. All men must therefore endeavor, "by force, or wiles, to master the persons of all men he can, so long, till he see no other power great enough to endanger him."

The result of this human nature, according to Hobbes, is a never-ending war of all against all. A life of perpetual warfare and destruction is obviously undesirable, but it is inevitable if human nature is given free reign. The only sensible alternative, Hobbes argued, is for all people to surrender their power to some common authority (the state) that is larger than any individual and that can and will "keep them all in awe," that is, keep human nature in check.

What Hobbes meant when he said that it was human nature to pursue the satisfaction of unlimited desires is that one could not be a person and act any other way. People had no more control over their natures and actions than falling rocks did over theirs. People were machines, set in motion by external, causal forces. They could be influenced by influencing the forces acting on them. But they could not exert self-restraint; restraint would have to be imposed from without, by a powerful state, the Leviathan.

Hobbes's views about human nature may seem a little extreme, but there is a distinctly modern ring to them. Hobbes regarded people as isolated individuals, with their own individual interests. It is the hallmark of modern, liberal democracy to regard people in just this way, as isolated individuals, in pursuit of their individual interests. Hobbes acknowledged that the precise nature of these interests would vary from person to person and from place to place, but he argued that everyone will have them and will do whatever can be done to satisfy them. Hobbes thought that people would be in constant competition, trying to outdo one another. Most of us certainly experience the reality of "keeping up with the Joneses"—that luxuries become necessities when everyone has them. We may differ from Hobbes in supposing that there are limits to what people will do. We may believe, unlike Hobbes, that even without a state policing the behavior of individuals, people will not indiscriminately try to steal from one another, that it is wrong for people to engage in ruthlessly aggressive competition, and that many people, if not all, will think twice before doing what is wrong. Hobbes might agree that it is wrong to pursue one's interests no matter what. But he would argue that good or bad, right or wrong, it is simply the way people are. Deceive yourself into thinking they are otherwise, and you will be swallowed up.

It is possible to agree that Hobbes's description of human nature has a ring of truth to it but put a very different interpretation on it than Hobbes did. One possible interpretation—the one Hobbes intended—is that people are selfish by natural law. They couldn't be, and never have been, otherwise. But another possible interpretation is that people are selfish only under certain conditions, conditions that could be different. It makes an enormous difference which of these interpretations one has in mind in claiming that it is human nature to be selfish. Consider two people, both of whom agree with Hobbes to the last detail that people are selfish, endlessly acquisitive, competitive, and aggressive. One person also believes that it is in the nature of the species to be this way. The other person attributes it to the peculiar conditions of a competitive, free-market capitalist society, in which every person is expected to fend for himself. The first person might, like Hobbes, seek some kind of social organization that could keep ruthless selfishness in check. The second person, unlike Hobbes, might seek some kind of social organization that would create people who were not so acquisitive, competitive, and aggressive. The first person might, again like Hobbes, see the need for an authoritarian state. The second might argue for some form of social democratic, welfare state. Thus two radically different prescriptions for social, economic, and political organization could arise from an agreed upon view of what people are like, if there is disagreement about why they are like that.

And views like the hypothetical ones of our anti-Hobbesian have almost as long a tradition as the Hobbesian view. The French philosopher Rousseau argued for what is sometimes called a "noble savage" view of human nature. On this view, people in the state of nature are cooperative rather than competitive. They show care and concern for their fellows and are altruistic rather than greedy. This benevolent innocence is corrupted by the social institutions within which the person grows, institutions that demand competitiveness and beat down cooperation.

The conflict between these two views has been played out often, in scholarly argument, in literature, and in late-night, beer hall conversation. It is a conflict whose resolution is enormously important, since it bears on the kind of society one aspires to create. But it is also a conflict whose resolution is enormously difficult. For both parties to the conflict agree about the facts—that people are selfish, competitive, and aggressive. They disagree about interpretation; incorrigibly evil human nature on the one hand; and corrigibly evil social institutions on the other. And it is not at all clear how to decide between these interpretations.

Biology, Behavior Theory, and Economics

This issue has the character of a chicken and egg problem. Which came first, competitive human nature or competitive society? Like the chicken and egg problem, it is easy to imagine clever, but interminable debate. It is hard to imagine resolution. But this discussion is about science, and science does not tolerate interminable debates that get nowhere. If there is to be a science of human nature, it will surely have to resolve this problem. What can be done to find evidence on one or the other side of the issue? The answer is provided in the work of Charles Darwin.

Darwin spent many years carefully observing the behavior of a wide range of animal species in their natural environments. Here, there was no chicken and egg problem. There was no culture or society to influence the desires and actions of snails, birds, and turtles. By watching them closely, one could get a picture of their nature. And then, perhaps, one could argue by analogy to human nature.

Darwin's observations led to *The Origin of Species,* in 1859. In it, he began with the view that life in the natural world was a battle among individual organisms for limited resources. Those individuals that won the battle and obtained the necessary resources would survive and reproduce. Those organisms that lost the battle would die off. The critical insight for Darwin was that not all members of a species were identical. Individuals differed in subtle but important ways. It was these individual variations that gave some members of a species a competitive advantage over others. The successful species members passed on their advantageous qualities to their offspring. The offspring thus also enjoyed a competitive advantage over other species members. In each generation, the organisms that possessed these superior qualities lived longer and had more offspring than organisms lacking these qualities. Over the course of many generations, more and more of the surviving offspring shared the qualities that had initially given a handful of individuals an advantage. Ultimately, these qualities became a nearly universal characteristic of the species. Thus the old species, which did not in general have these qualities, would evolve into a new species that did. And the process of evolution was a continuing one: species change, over the course of thousands of years, was the way nature worked and would continue to work. Darwin's theory contained two crucial elements. First, there was *variation* among members of a species. Second, a process of *natural selection* seemed to increase over generations the number of members of a species with characteristics that increased their chances of survival.

Darwin's theory had a major influence on all aspects of the intellectual world. With respect to developing views of human nature, the theory of natural selection suggested that there was continuity among species, that the prevailing dichotomy between "man and beast" was inappropriate. The notion of species continuity made it easier to argue against the prevailing wisdom that people were in all important respects unique. It made it easier to justify looking at what animals were like in nature to find out what people *would* be like in nature. And it firmly grounded as a universal principle throughout the natural world Hobbes's picture of a competitive struggle for survival. What separated people from beasts was not that people were above the struggle, but that they could create social institutions that would prevent them from struggling to the death.

The modern incarnation of this Darwinian idea is to be found in evolutionary biology, especially a branch known as sociobiology. The central dogma of evolutionary biology is that significant characteristics are passed from parents to offspring in the genes. Not all organisms will be successful at surviving and reproducing, and thus not all genes will be equally likely to pass from one generation to the next. The genes that do survive will be the ones that make successful organisms. Thus, only organisms that successfully satisfy their interests will contribute their genes to future generations. As a result, only genes that see to the successful pursuit of self-interest by the organisms possessing them will survive. The implication of this line of thinking is that selfishness—the single-minded pursuit of genetic self-interest—is a biological fact of life, a natural necessity. This is the worldview of modern evolutionary biology: selfish organisms whose behavior is largely genetically determined and whose mission in life is the maximization of reproductive success.

Developing alongside evolutionary biology there is another discipline, also cast in the shadow of Darwin. It is a branch of psychology known as behavior theory. Its current form is most closely associated with B. F. Skinner, but the origins of its central ideas can be found in Herbert Spencer, an approximate contemporary of Darwin's who was greatly influenced by Darwin's theory. Spencer applied Darwin's theory of variation and selection to the life history of individual organisms. He suggested that organisms engaged in essentially random activity. Some of that activity resulted in pleasurable consequences, in the satisfaction of desires. These pleasurable consequences worked to select the activities that preceded them. Activities that either failed to produce pleasurable consequences or, indeed, produced unpleasant ones would stop occurring. Over the course of its life, this process of selection would yield an organism whose repertoire of actions was effective in meeting its desires. This process of selection came later to be called *the law of effect*. If we make the not implausible assumption that the things an organism desires are also things that promote its survival, the law of effect works hand in glove with evolution. Natural selection across generations selects those traits that best promote survival, and the law of effect within generations does the same.

Both behavior theory and evolutionary biology focus on the individual organism as the critical unit of analysis. Organisms are selfish, out to maximize their interests. In the case of evolutionary theory, what is being maximized is reproductive fitness. In the case of behavior theory, what is being maximized is pleasurable events. The pursuit of self-interest is nature's way of improving both individuals and the species as a whole. In fact, the notion that pursuit of self-interest was nature's way was used very early to justify and defend a kind of laissez-faire social organization in which every individual was expected to fend for himself. The argument made by so-called social Darwinists was that various forms of social welfare and aid to the weak went against nature. Natural competition selected the strong, the fit. If society held out support for the unfit, it would keep bad genes, or ineffective behavior, from dying out. In the long run, this short-sighted charity would work to the detriment of the human species. The weak weren't meant to survive; they weren't meant to inherit the earth. The social Darwinist movement, and its use of Darwin's theory as a *natural* defense of social policy, has been well documented by Richard Hofstadter in his book, *Social Darwinism in American Thought*. Hofstadter says:

> In the Spencerian intellectual atmosphere of the 1870's and 1880's it was natural for conservatives to see the economic contest in competitive society as a reflection of the struggle in the animal world. It was easy to argue by analogy from natural selection of fitter organisms to social selection of fitter men, from organic forms with superior adaptability to citizens with a greater store of economic virtues. . . . The progress of civilization . . . depends on the selection process; and that in turn depends upon the workings of unrestricted competition. Competition is a law of nature which "can no more be done away with than gravitation," and which men can ignore only to their sorrow.

What is especially significant about this movement is that it was an attempt to take issues that were traditionally regarded as moral issues—issues of right and wrong, and of duty and obligation—and turn them into scientific issues—issues of fact. Whether society regarded it as moral or immoral to let the weak perish was beside the point. It was nature's way. As one social Darwinist, William Graham Sumner, put it, "The truth is that the social order is fixed by laws of nature precisely analogous to those of the physical order."

Thus a branch of biology, evolutionary biology, and a branch of psychology, behavior theory, are converging on a science of human nature that is the fleshing out of Hobbes's speculation. But the jigsaw puzzle is not complete. There is a third piece, a third perspective on human nature that meshes with the first two. Historically, it is the oldest of the three, and in most people's lives in contemporary society, it is the most prominent and influential. It is the picture of human nature purveyed by modern economics. Beginning with Adam Smith, in the eighteenth century, modern economics has claimed that it is human nature "to truck and barter," to seek profit through exchange, to "buy in the cheapest and sell in the dearest market." People have wants, unlimited wants, and their mission in life is to pursue these wants to the hilt. Man is "economic man," out to maximize self-interest. A competitive, free-market economy allows people to express their basic nature. It is the only form of economic and social organization that is in harmony with man's basic biology. According to Smith in his book, *The Wealth of Nations,* a competitive free market is best not because people *ought* to be free to buy and sell as they please but because any other form of economic organization will do violence to man's basic nature. The edifice of modern, industrial capitalism that is so familiar is built upon Smith's "economic man" foundations. It is built upon the view that one could as well do away with the competitive pursuit of self-interest as one could do away with gravity. Again, Hofstadter says it well:

> A parallel can be drawn between the patterns of natural selection and classical economics. . . . Both assumed the fundamentally self-interested animal pursuing, in the classical pattern, pleasure or, in the Darwinian pattern, survival. Both assumed the normality of competition in the exercise of the hedonistic, or survival impulse; and in both it was the "fittest" . . . who survived or prospered—either the organism most satisfactorily adapted to its environment, or the most efficient and economic producer, the most frugal and temperate worker.

These three disciplines—economics, evolutionary biology, and behavior theory—are all concerned with developing a picture of human nature, a picture that conforms to the canons of scientific methodology and satisfies the tough standards of scientific rigor. Traditionally, each of the disciplines has acted as though it could do the job alone. Each has its especially zealous practitioners who claim that a comprehensive theory of human nature can be produced strictly within the confines of his or her particular discipline. We will see that while each of these disciplines has a claim to be taken seriously, none has

a claim to comprehensiveness. We will also see that rather than presenting three different theories of human nature, these disciplines
are really presenting the same theory, but from slightly different per-
spectives. When taken together, they constitute a truly general theory of what it means to be a person. And it is a theory that is of a
piece with the views of Thomas Hobbes.

"Is," "Ought," and Science

This chapter began by posing some moral issues that have per-
plexed people for millennia and by introducing Hume's famous distinction between "is" and "ought," and his claim that conclusions
about the way things should be cannot be derived from any knowledge
about the way things are. The domain of "is" is the domain of facts,
and it properly encompasses the whole of the natural and social world.
The domain of "ought" is the domain of morals, and its compass is
restricted to certain aspects of human action. Only human beings are
concerned with what ought to be. Nevertheless, the facts—what is
the case—can be relevant to morals—what ought to be the case—
even if they don't imply them. Conceptions of what people are and
can be constrain how people reason about what they ought to be.

Most people are quite comfortable thinking about the world in this
way. Facts are one thing; morals are another. Facts can be ascer-
tained unambiguously; morals are matters of opinion and dispute.
Facts are the province of scientific experts; everyone is equally expert
(or inexpert) about morals. But not so fast. We shouldn't get too
comfortable. We may have settled into dividing the world into natural
and moral categories, but about which aspects of human activity belong
in which category there is continuing dispute. As society's dominant
conception of human nature comes increasingly to approximate the
scientific one offered by evolutionary biology, economics, and behavior theory, there is persistent pressure to appropriate more and more
aspects of human life to the natural.

How can issues that are now regarded as unambiguously moral be
relocated in the domain of facts? Well, think for a moment about
what makes something a moral issue. Abortion, capital punishment,
marital fidelity, aggression, economic organization, and the like are
regarded as moral issues in part because people believe that they are
all subject to human discretion and control. After all, not all of these
activities themselves are uniquely human. Other animals mate, raise
children, allocate resources, and fight. It's just that most people believe

that while these other animals can't help what they do, human beings can. If people couldn't help it, arguments could probably continue about right and wrong, but they would lose much of their significance. Well, the scientific conception of human nature suggests that much of what is regarded as discretionary about human action really is not. Instead, there is a natural necessity that impels people to act in the ways that they do.

When someone asks why Mrs. Jones held three jobs to put her kids through school, the immediate answer is that it's only human nature for parents to be devoted to their children. When someone asks why a lost wallet was returned anonymously to its owner with all the money gone, the reply is that it's only human nature for people to take something if they think they can get away with it. When someone asks why Mr. Smith, who was just convicted of tax evasion, would cheat on his taxes when he has enough money to last him a thousand years, the reply is that it's only human nature for people to want to keep what they've earned, no matter how little they may need it. It's only human nature this, it's only human nature that—again and again we read or hear this sort of explanation; Mr. X did Y because it's only human nature to do Y.

"It's human nature," then, is a shorthand way of saying that a particular action is perfectly consistent with common rules-of-thumb about what people are and how they act. It sometimes goes disguised as "what's the big deal; what else did you expect." It sometimes goes disguised as a savvy, knowing smirk. Most often, though, it goes disguised as invisible; like Sherlock Holmes' dog that *didn't* bark in the night, "it's human nature" is the answer to the questions that aren't asked. But what is it that is really being said when some action is explained by appealing to "human nature"?

Let's consider some possible different meanings one might have in mind in saying that "it's human nature". For concreteness, consider the claim that it's human nature to be selfish. This claim might mean:

1. It is the way all people everywhere are and have always been, and must necessarily be. It is part of the essence of being a person. A person who wasn't selfish would be like a bird without wings.

2. It is the way all people everywhere are and have always been and must necessarily be under certain natural conditions. For example, people always act selfishly when their lives are at stake, although they might not at other times.

3. It is the way some, most, or all people are and are likely to

continue to be under a particular set of social, economic, and cultural circumstances that could be otherwise, may once have been otherwise, and may still be otherwise in different cultures. For example, people who grow up as part of the modern "me generation" are selfish.

Now the chances are pretty good that someone who says "it's human nature" hasn't given a great deal of thought to which of these meanings he has in mind. However, in terms of the potential consequences of appealing to "human nature" as an explanation, it makes a big difference which of these meanings is implied. Suppose we agree that selfishness is a bad thing, that people should not be selfish. Unfortunately, though, it's human nature to be selfish. What can be done about it? If it's human nature, meaning 1, there is really nothing to be done. If part of what it means and has always meant and must always mean to be a person is to be selfish, then we simply have to resign ourselves to the fact that people possess an ineradicable character flaw. Social institutions can be designed to restrain people from exercising their selfishness. People can be policed and severely punished for displays of selfishness that are especially abhorrent. Or perhaps, in light of the facts, the moral view that selfishness is bad can be revised. But not much can be done to change selfishness.

If selfishness is human nature, meaning 2, there is reason to be a little more hopeful. While there isn't much to be done to prevent selfishness from appearing when the natural conditions that give rise to it occur (when, for example, people's very lives are being threatened), things can be done to keep those natural conditions from arising. By keeping situations in which people are at risk at a minimum (by guaranteeing, for example, adequate food, housing, and medical care to everyone), outbreaks of selfishness can be controlled.

If, finally, selfishness is human nature—meaning 3, there is cause for real optimism. If it is human nature to be selfish under a particular set of social conditions, then selfishness can be eliminated just by changing the conditions. Thus, it makes a big difference what meaning underlies the claim that such and such is human nature. Two people could agree that it is human nature to be selfish. This view might lead one of them, operating under meaning 1, to a life of cynicism and suspicion. It might lead the other, operating under meaning 3, to a life of commitment to producing social change.

It is a truism that knowledge is power. Like most truisms, this one is not entirely true. Knowing what forces are responsible for keeping the planets moving around the sun does not give people any particular power to control or change them. Knowing that the sun is dying and that the universe is expanding does not give people any particular power to prevent the eventual termination of life on earth. Knowing that it is human nature (meaning 1) to be selfish would not give people any particular power to control or change it. On the other hand, knowing it is human nature (meaning 3) to be selfish *might* give people the power to control and change it.

It is important to realize that what meaning of "it's human nature" we have in mind when we use that phrase to explain some action should affect how seriously we take the explanation. "It's human nature"—meaning 1, is pretty weighty. To be told this is to be told something fundamental, with major implications. "It's human nature"—meaning 3, is much less weighty. It is tempting to reply to this sort of claim, "So what. People shouldn't be acting in this way, and if these conditions are making them, then they had better be changed." If some human characteristic is human nature—meaning 3—then people retain control and discretion in the exercise of that characteristic by virtue of their control of the conditions that foster it. If, in contrast, some human characteristic is human nature—meaning 1—then people have neither control nor discretion in the exercise of that characteristic. They just can't help being selfish. No one could.

Indeed, the sense of human nature conveyed by meaning 3 is as something that is mutable rather than fixed. Selfishness is in human nature in the same way that the David was in the marble before Michaelangelo touched it. In an important sense, the David *was* in the marble; with a different block of marble, it would have come out looking different. But lots of other possible statues were also in the marble, and what actually came out was the product of the conditions to which the marble was subjected (Michaelangelo's vision and his hands). Based on meaning 3, selfishness is in the human marble, just requiring the right conditions to make its appearance. But so, for all we know, is selflessness. So are lust and chastity, greed and generosity, deceit and honesty. Clearly, though the David was in the marble, we are less interested in the marble than in the conditions that shaped it. And if selfishness is in people in the same sense, we should here, too, be less interested in the marble than in the conditions that shape it.

An appeal to human nature, meaning 3, preserves the significance of moral arguments. If it is human nature to be selfish only under certain conditions that can be controlled, there is plenty of room for discussion of whether those conditions should or should not obtain. On the other hand, an appeal to human nature, meaning 1, obviates much of the discussion of the moral significance of selfishness.

It is the hallmark of science that it seeks explanations that have the character of meaning 1. It seeks the eternal regularities of things, the essential natures of planets, bees, and people. It is in the nature of being a person to need food to eat and oxygen to breathe; it is in the nature of being a person to see only certain wavelengths of light and to hear only certain frequencies of vibration of air. Moral and immoral are beside the point. What characterizes the disciplines of economics, evolutionary biology, and behavior theory is that they are endeavoring to extend explanations that have the character of meaning 1 to domains of human activity that are presently regarded as discretionary—domains involving the pursuit and distribution of resources, patterns of mating and child care, patterns of aggression and cooperation, and patterns of social organization. If they are successful, it may become as idle to talk about the morality of selfishness as it is to talk about the morality of planetary motion.

Why should the encroachment of morality by science concern us? These disciplines are making substantive, factual claims about human nature. They are scientific claims, and like all scientific claims, they must survive empirical test. If the claims are false, if they fail the test, then this science of human nature will die, of "natural" causes. If the claims are true, on the other hand, there isn't much to be done about them. So why worry? What's the big deal?

The rest of this book can be seen as an attempt to provide an answer to this question. I believe that the conception of human nature shared by economics, biology, and behavior theory is false as a universal. That is, it is not human nature—meaning 1. It is, however, human nature at certain times and places, human nature—meaning 3. Moreover, the conditions under which it is human nature presently obtain in Western, industrialized society, and ideas from economics, biology, and psychology have helped to bring these conditions into existence and presently help to justify their continued existence. In short, these disciplines provide a partly accurate description of human nature, but of a human nature they have helped create. Finally, whether the conditions that promote this kind of human nature are present or not is a matter of discretion, of human decision and control. Whether we want them depends on what we think people should be, on how we think people should act. Thus, our judgment of the sciences of economics, biology, and behavior theory must be, in part, a moral one. Criticism of these disciplines is in part an attempt to rescue the oughts of human nature from encroachment by the domain of is.

And each of these disciplines is vulnerable to serious criticism. They do not, either in isolation or taken together, warrant our unqualified allegiance. Yet we seem, perhaps unwittingly, to be giving them increasing portions of our allegiance every day. When we examine aspects of our everyday life and of the social institutions that contribute to it, we find places where this Hobbesian, scientific view of human nature has already had a tremendous impact, as well as places where it may have an impact in the future. It is an impact that has not been benign. Many of the features of modern life outlined in the last chapter that we find so distressing can be traced, in part, to an underlying commitment to the Hobbesian picture, buttressed by the claims of these three modern sciences.

3

Economics and Human Nature

I have come to believe that the economic approach is a comprehensive one that is applicable to all human behavior, be it behavior involving money prices or imputed shadow prices, repeated or infrequent decisions, large or minor decisions, emotional or mechanical ends, rich or poor persons, men or women, adults or children, brilliant or stupid persons, patients or therapists, businessmen or politicians, teachers or students.

GARY BECKER

The first principle of economics is that every agent is actuated only by self-interest.

FRANCIS EDGEWORTH

In 1650, a Boston minister gave a sermon in which he enumerated the immoral principles of trade. First among them was this: that a man might sell as dear as he can, and buy as cheap as he can. Less than two hundred years later, all of Europe and the United States was operating in accord with that very principle, not because it had been newly judged to be moral, but because it had been judged to be human nature. The midwife of this transformation was Adam Smith. In *The Wealth of Nations*, published in 1776, Smith put forth a conception of our economic nature—of what has come to be called "economic man." That conception, and the principles of social organization that seemed to follow from it, have guided the discipline of economics, and many of us, ever since.

Economics is principally concerned with the allocation of scarce resources. "Scarce," here, doesn't have to mean rare. A resource is scarce if people seem to want more of it than there is to go around. Under these conditions, some sort of scheme is required to determine who gets what. Economics asks two different sorts of questions about scarcity. First, it asks what will happen if one or another scheme of resource allocation is adopted. That is, it tries to *predict* the effects of different kinds of distribution schemes. Second, it asks whether some distribution schemes are better than others. That is, it tries to determine whether some system for allocating resources can be found that is more efficient and effective at satisfying people than any other.

We can see these different characteristics of economics by looking at an example. What is the economics of breathable air? Until recently, there was no economics of breathable air, despite the essential role it played in sustaining human life. That is because it was not a scarce resource. There was plenty of it to go around, more than enough for everybody. So there was no need to ask questions about allocation schemes. Indeed, there was no need to have allocation schemes. Unfortunately, as those who live in urban, industrial settings can attest, that is no longer true. Breathable air is now a scarce resource. Enter economics. The economist can set out to determine what will happen to the supply and distribution of breathable air if a variety of different measures are adopted. What will happen if oxygen meters are put on people and they are required to pay for what they use? Or what if everyone is given a certain minimum amount of air for free and people are only made to pay for what they use beyond this minimum? Or what if air stays free, but laws are passed against pollution that will increase the production costs of many industries? Or suppose instead of being prohibited from polluting, polluting industries are required to pay a pollution tax? And what will happen if nothing at all is done, allowing people who can afford it to buy breathable air in the form of vacation houses in the country? Finally, is there some scheme for the allocation of breathable air that is superior to all others in satisfying the members of society?

Now while the discipline of economics is less than three hundred years old, resource scarcity is considerably older. There has probably never been a time in human history when at least some resources weren't scarce. Before there was economics, societies dealt with scarcity by appealing to tradition, or perhaps to moral considerations. A society might give the elders, or the men, first crack at scarce resources.

It might apportion resources in accordance with some measure of social standing. It might share all resources equally among its members, or set up some rules of ownership. It might decide to look after the weak or helpless and let all others fend for themselves. There are countless possible schemes for resource allocation, some based on accidents of history and others based on considerations of fairness or justice. What economics has done is try to replace moral and traditional principles of resource allocation with *scientific* ones. Economics is the science that attempts to determine what the consequences of various allocation schemes will be. It is the science that tells society how to organize itself to meet best whatever allocation goals its citizens have.

As a science, economics starts out with certain presuppositions about human nature—about what people are and what they want. The presuppositions are quite simple. People, by their natures, are motivated to pursue their own interests. When given a choice, people will select an alternative that serves their individual interests. Indeed, more specifically, people are motivated to *maximize* their self-interest, and in acting and choosing, they behave in a way that is consistent with the maximization of self-interest. Economists termed such pursuit of self-interest "economic rationality," and human nature as economic rationality is where economics starts. From here it can go on to ask how individual, rational economic agents will and should act under one or another set of economic conditions.

An economy, in this view, is a collection of rational economic agents. Often, they are organized into groups. A family is one such group. A firm or business is another. Families act to maximize their interests, as do firms. The actual interests of each group may differ, as may the interests of families in general from firms in general, but whatever these various interests are, we can be sure that each group is motivated to maximize them. And when all these different groups in pursuit of self-interest are aggregated, they make up an economic system. An economy as a whole is nothing but the economic activity of its individual members.

This simple conception of human nature provides the foundation for economic science. This chapter will explore where this conception came from, how it gave rise to an economic science, and what modern economic science looks like.

The Origins of Economic Science

We can begin to see where economics came from by looking at the word *economics*. It derives from the ancient Greek word *oeconomics,* which refered to principles of management of the household. And this meaning was preserved until well into the eighteenth century, to the time of Adam Smith, as evidenced by the fact that Smith's teacher, Francis Hutcheson, wrote a book called *Short Introduction to Moral Philosophy,* in which the section on "principles of oeconomics and politics" had chapters on marriage and divorce and on the duties of parents and children. Thus, economics seemed to encompass all aspects of the production of daily life, from putting bread on the table to raising children to be responsible citizens. With this sense of "economic," all people could be thought of as "economic men."

And how should "economic men" act? They should act "rationally." Again, a look at the origins of a word is instructive. To the Greeks, "rational" was "proportional," not altogether surprising since the root of "rational" is "ratio." So to act rationally was to keep things in proportion, to balance the various aspects of one's life and organize them into a harmonious whole. The rational person devoted the right amounts of time and resources to his work, to his family, to study, to recreation, and so on. The rational "economic person" was the one who kept all the aspects of the production of daily life in proportion.

This was the conception of "economic man" that Adam Smith inherited. But in *The Wealth of Nations* he transformed it, and the world has not been the same since. First, Smith suggested that economics was not about the production of daily life by the household. It was about the *exchange,* the trading among people of goods and services. How this loaf of bread or that pair of shoes came to be, or came to belong to a particular person, was not an economic question. The economic question was what it would take for that person to part with it. For what would she give it up? What would induce her to offer many loaves of bread, or pairs of shoes, to others? So principles of economics were to be principles of exchange, principles that described people's natural "propensity to truck and barter," to trade one thing for another.

Second, an economy was made up of people whose behavior was governed by these principles of exchange. It could be looked at as an autonomous system, governed by its own laws, and independent of other social institutions like the church, the family, and the state. To understand an economic system, one did not need to know every-

thing there was to know about a society. Societies and the people in
them were multifaceted to be sure. But the economic facet, of both
individuals and societies, could be understood apart from everything
else. So the way in which one got along with one's spouse, or raised
one's children, though obviously important, were no longer to be
thought of as part of economics.

The stage on which economic actors played their parts was the
market. The market was where the exchange occurred, where people
trucked and bartered. The market was where a person became an
"economic man." And what did an economic man do? First and fore-
most, economic man pursued self-interest. He was out to maximize
gain, by buying as cheap as he could and selling as dear as he could.
He was greedy, possessing unlimited wants. Indeed, there was no
principled distinction to be made between wants on the one hand and
needs on the other (as we might casually distinguish between wanting a Mercedes and needing food). The reason for this nondistinction
is deep. To talk about human needs, one must have a theory of value—
of what is important and why. For Smith, there was no theory of
value as intrinsic to acts or commodities. Value, at least *economic*
value, was to be determined in the marketplace, by the prices commodities would fetch, by what the traffic would bear. Notice the
qualifier "economic" value. Even without a developed idea of what is
valuable to people and why, people can still readily agree that certain
things are valuable. Oxygen is valuable. A beautiful sunset is valuable. A loving relation to another person is valuable. These things all
have value, but not *economic* value. The reason for this is that they
are not subject to exchange in the marketplace. They may all have
use value, but they don't have *exchange* value. And economic value is
exchange value. That things of value may lie outside the sphere of
economics posed no special problem for Smith. Recall that he
acknowledged that the economic facet of people was only one facet
among many, albeit an autonomous and important one.

If the market is the stage for economic action, what does that action
look like? Well, suppose one day a father notices his young child
playing with an old bicycle tire. She puts it around her waist and
starts rocking back and forth to make it spin and keep it from falling.
Her delight as she plays with this makeshift toy is almost uncontrollable. "I've got it," her father says, "let's make them out of plastic and
call them hula hoops." So he does, and he brings them to the marketplace. People snap them up like wildfire, and since her father
gains or profits from each exchange (or else, why would he do it?), he

soon becomes a rich man. Others watch him shrewdly and decide to get in on a good thing. Soon, everyone is trying to sell hula hoops in the marketplace. No one is selling bread, or meat, or clothing. The best source of gain through exchange this year seems to be hula hoops. And so society slowly hula hoops itself to death from starvation and exposure.

This is hardly a satisfactory outcome of exchange in the marketplace. If everyone is going to be out to maximize self-interest in the market, someone is going to have to see to it that the needs of society are met. There will have to be some good samaritans. Or perhaps the state will have to intrude on the autonomy of the economy to make sure that basic goods and services are being provided for its citizens. Otherwise, it seems, all will be chaos, with wild swings in what is available in the market from day to day as people try to maximize gain.

Not so, said Smith. The beauty of the marketplace is that if it is left alone to operate in accord with the laws of economic activity, it will end up satisfying everyone's wants perfectly. The market will provide just the right amount of each and every type of good and service to satisfy everyone's desires. How can this be? Who can be smart enough and powerful enough to steer the market in this wonderfully efficient way? Can people be counted on to be benevolent enough, altruistic enough, to forsake a little bit of gain to provide for the needs of society? The answer is that they don't have to be. Smith tells us, "It is not from the benevolence of the butcher, the brewer, or the baker that we expect our dinner, but from their regard to their self-interest." The market is guided by an *invisible hand,* a natural and inevitable process that guarantees efficient results of exchanges. It is the hand of nature, the same one that keeps pushing the planets around the sun.

How can a bunch of unorganized individuals, each out for himself in a chaotic market, end up acting for the benefit of all? Smith's answer is that this will happen if the market is *free and competitive.* If anyone can sell whatever he wants, for whatever price he wants, at whatever time he wants, the competitive market will yield the right stuff, at the right time, for the right price. The magical mechanism of competition is both simple and beautiful. Here is how it works. The girl's father starts selling hula hoops and making a lot of money. Others decide to make and sell hula hoops too. Suddenly, there are lots of hula hoops available from different sources in the market. Indeed, the market may be glutted with hula hoops. There may be more of them available than people want to buy. So as to induce people to buy his, the father lowers his price (assuming, of course, that economic people will buy as cheap as they can). Others follow suit, lowering their prices. Some find that they can not sell hula hoops profitably at these low prices. They stop making hula hoops and make something else instead. The others continue to compete, driving each others prices down. So as to continue to make a profit, hula hoop makers find cheaper, more efficient ways to produce them. If still there are more hula hoops available than people want, even at these lower prices, other hoop makers will go out of business. The process will settle down into an equilibrium when the supply of hula hoops at a given price exactly matches the demand for them at that price. In economic jargon, the market for hula hoops will *clear*. If demand for hula hoops should increase, production will increase to match it. If it should decrease, more suppliers will stop making them.

These ex-hula hoop manufacturers will start bringing something else to the market; perhaps meat, or bread, or clothing. And the number of people supplying these various things will also be regulated by the demand for these things. In other words, with free competition, all markets will clear. All goods and services will be produced in quantities that match the demand for them at prices that are acceptable. What ensures the production of the right amount is the fact that there is profit to be had by meeting unsatisfied demand and loss to be incurred by overdoing it. What ensures a fair price is that there is profit to be had by anyone who can undersell the competition.

The beauty of this system comes from its inevitability, from its nondependence on charity, self-sacrifice, or good will. Competition will keep prices from being arbitrarily high. Producers will produce what consumers want, for if they don't, consumers won't buy. Thus, the notion of "consumer sovereignty" is born. Each person, as an actor on the economic stage, exerts control over what is available for what price by exercising his freedom to buy or refrain from buying. Consumers drive the marketplace. They get the goods and services they want, and thus they get what they deserve.

The invisible hand of the market works in this way for all commodities that the market offers. But there is one particular commodity that is so important it warrants special discussion. That commodity is labor. Just as bread, clothing, and hula hoops are trucked and bartered in the marketplace, so are the time and energy of people. The labor market is like any other market. When labor power is scarce, relative to the demand for it, workers can sell their time for high

prices. Indeed, the high price that labor brings may fetch new participants into the labor market. But as a result, supply will start matching demand, and may even come to exceed it. When this happens,
laborers, to get work, will have to start offering bargains; they will
have to lower the price for their services, or else be driven out of the
market by the competition. Like all other markets, the labor market
should clear. A point should be reached at which the supply of labor
at a given price exactly matches the market's demand for it. What
this means is that there is no such thing as involuntary unemployment. The unemployed laborer can always get a job by offering to sell
his time for less than the going market rate.

This, then, is Adam Smith's story about how selfish individuals,
out only for themselves, will act in a manner that is good for everyone. The free market mechanism ensures a favorable result for society. But the market will not function in this efficient manner under
all circumstances. A few conditions must be met in order for the
market to do the job that Smith claimed it could. First, it must be
free. People must be allowed to make the trades they want, unfettered by restraints or prohibitions. Prices must be allowed to vary
freely, to be determined by the interplay of supply and demand. For
it is the effect of supply and demand on prices that is the principal
engine for assuring the availability of just those goods and services
that people want, at the prices they are willing to pay. Second, there
must be competition. There must be many people offering the same
goods and services. It is their competition for the consumer's dollar
that will make price sensitive to demand. If only one person sold
bread in the marketplace, he could get away with charging almost
anything. Or if various bakers worked in collusion, they could fix
prices artificially high, or create scarcity by baking less than people
wanted, with the result that prices would get high naturally, as buyers competed for the bread that was available. Thus, a market dominated by one producer—what is now called a monopoly—will not
work in the salutory way that Smith envisioned.

Furthermore, buyers in the marketplace should not care who they
buy from. They should be out to make the best deal they possibly
can, not to help out their brother-in-law, or patronize someone who
is a nice guy, or buy from the person whose political views match
their own. The market is predicated on the idea that people enter it
wanting *only* to buy as cheap as they can and sell as dear as they can.
If they have other concerns that affect their behavior in the marketplace, these concerns will interfere with the interplay of supply and

demand in determining prices. Another way of making this point is to say that sellers in the marketplace should be essentially anonymous and interchangeable. All that should matter is product and price. To see why this is important, consider an example. Suppose there are two principal manufacturers of chocolate chips. They compete with each other, and thus prices are kept down. Now further suppose that for political reasons, a massive boycott is organized against one of the companies. People refuse to buy their chocolate chips no matter how good the product or how low the price. What this does is wipe out the market as a regulator of prices. The other company can now start raising its prices with impunity, knowing that the boycott will keep people from purchasing the chips of their now cheaper competitor.

Similar conditions must be met for the labor market to function properly. Workers must be free to choose the jobs they want, and bosses must be free to pay whatever supply and demand allow them to get away with. Closed unions interfere with the former, and minimum wage laws interfere with the latter. There must be competition among individual workers. Again, labor unions diminish competition. And there must be anonymity. Only product (skill) and price (wage) should determine the purchase of labor. If other factors, like political affiliation or skin color, enter into labor purchase decisions, the market in labor will not function properly.

Even when these conditions are met, there will be problems in the functioning of the market. It takes time to switch from making hula hoops to making bread, and even if the market assures that society will have just the right number of each kind of producer, those who are forced to shift will surely suffer hardships during the transition. The same is true of laborers who are forced to switch jobs. It takes time to learn new skills, one may have to relocate, and so on. The *ideal* description of the market assumes that responses to changes in supply and demand occur instantaneously, when, of course, they do not. But if we ignore for the moment the practical fact that adjustments to changes in the market take time, and look only at the effects of a market system on society as a whole, we can appreciate Smith's optimism that free markets were the way for societies to get what they wanted most efficiently.

And there is more to the story than this. The economic clockwork is not stagnant. It has an engine for the growth of material well-being built into it. The engine is technology. Producers will make profits; they will accumulate capital. This accumulated capital they will invest, in the hope of making still greater profit. Investment will be directed toward making things better, cheaper, and faster, to give producers and edge on the competition. It will be directed to technological advance, to increased specialization and automation that will increase the overall productivity of society, making everyone better off. This technological advance will depend on the accumulation of capital by a few especially successful people of commerce, for it would be far too costly for small producers to contemplate. Thus, though unequal distribution of resources may be a natural consequence of the free-market system, it will have salutory effects for everyone. The rich may get richer, but the poor will get richer, too. One may recognize this line of argument, in modern form, in the "trickle-down" theory used by the Reagan administration to justify decreasing rates of taxation of businesses and of the wealthy. The idea is that everyone will benefit from an increase in investment by those with the resources to invest.

Smith delivered a now famous paean to the virtues of technological advance and specialization of labor in his description of a pin factory:

> One man draws out the wire, another straits it, a third cuts it, a fourth points it, a fifth grinds it at the top for receiving the head. . . . I have seen a small manufactory of this kind where ten men only were employed. . . . They could make among them upwards of forty-eight thousand pins a day. . . . But if they had all wrought separately and independently . . . they certainly could not, each of them, make twenty.

Here, in 1776, is a forecast for a technologically advanced, prosperous society, in which every day some item that once was a luxury for the few becomes available to the many. Furthermore, it is a society that, at least in its commercial aspects, is thoroughly democratic. All people are, in principle, equal in their capacity to bend the market to their interests and desires. There is no authority with the power to tell anyone what to buy or sell. In the marketplace, each person is sovereign. And the magic that makes this all possible is the simple fact about human nature that people, if left alone, will pursue their rational self-interest. The rationality of selfishness turns out to be good for everyone.

Smith's story should give us pause. On the one hand, it is such an accurate description of modern industrial society that it is hard to believe that Smith wasn't writing yesterday. One can find someone making a Smithian defense of the free market almost daily in newspapers and magazines. But on the other hand, it is a little odd to see the pursuit of self-interest extolled in this way. While almost everyone acknowledges it as a fact of life, very few people revere it. It is generally regarded as a necessary evil, a human tendency that must be restrained, or harnessed productively. The immensely successful entrepreneur earns grudging respect, but his profit seeking is regarded as a means to higher ends, not as an end in itself. People admire self-sacrifice, not greed. Yet Smith seems to view the unlimited greed that drives the accumulation of wealth and subsequent growth of technology as an unqualified good, as essential to the continuing prosperity of a society. Could this be right? If so, what is to stop people from doing absolutely anything in the service of gains in the marketplace? What is to stop people from sabotaging—even killing—their competitors? What is to stop them from cheating their customers? What prevents the market from becoming a jungle?

The answer to these questions is that Smith did not regard unbridled greed as an unqualified good. It was the way people were—and *should* be—in the economic sphere of their lives. But economic man, remember, was just one piece of man—one piece among many. There were other pieces, and these pieces reflected what people regard as the nobler aspects of human nature. They also served to keep the economic piece in check, to keep people from doing anything whatsoever in the service of gain. These other pieces of human nature were moral, and they were reflected in a person's concern for his fellows. Owing to what Smith called "natural sympathy," people were able to see themselves in another's place and to act accordingly. Natural sympathy had an important role to play in the market; it restrained the greedy pursuit of self-interest that might otherwise set people against one another. In *The Theory of Moral Sentiments*, Smith said:

> All the members of human society stand in need of each other's assistance. . . . Where the necessary assistance is reciprocally afforded from love, from gratitude, from friendship and esteem, the society flourishes and is happy. All the different members of it are bound together by the agreeable bands of love and affection. . . . Society, however, cannot subsist among those who are at all times ready to hurt and injure one another.

Thus, love, gratitude, friendship, and esteem, none of them commodities one can buy cheap and sell dear, are the essential glue that holds a society together. And for a market to function, society must be held together. Since these unselfish characteristics are as much a part of people as is the pursuit of self-interest, there is little reason for concern that the market will turn people against one another.

Nevertheless, Smith had concern, and he expressed it. His worry was that habits of behavior that were developed in the market might have spillover effects outside it. For example, he saw a dark side to the very specialization of labor that would inevitably enrich society's material well-being. He feared that the repetitiveness of the pin factory might be intellectually deadening:

> In the progress of the division of labour, the employment of the . . . great body of the people comes to be confined to a few very simple operations, frequently to one or two. But the understandings of the greater part of men are formed by their ordinary employments. The man whose life is spent in performing a few simple operations has no occasion to exert his understanding, or to exercise his invention in finding out expedients for difficulties which never occur. He naturally loses, therefore, the habit of such exertion and generally becomes as stupid and ignorant as it is possible for a human creature to become.

So even if the market couldn't do everything for society, and even if economic rationality was not the whole story about people, and even if there was a danger that aspects of market behavior might spill into domains of life where they didn't belong, for what the market was intended to accomplish, Smith was convinced that it couldn't be beaten. Far from being a sin, the quest for profit was man's salvation.

The Science of Economic Rationality

Adam Smith was the father of modern economics. We have seen that for him, economic man, guided by self-interest, and moral man, guided by sympathy, could coexist. As his ideas were developed, over the next two centuries, the notion of economic man was increasingly refined and elaborated. At the same time, the notion of moral man slowly disappeared, at least from economics. The reason for this shift in emphasis may have been strictly tactical. Economic science was to be *science*. This meant that it had to be quantified; it had to be precise. But certain aspects of human nature seemed to resist quantification, at least for the time being.

This strategic decision is reflected in the words of Alfred Marshall, in the eight edition of his book, *Principles of Economics,* published in 1920:

> Such a discussion of demand as is possible at this stage of our work, must be confined to an elementary analysis of an almost purely formal

kind. The higher study of consumption must come after, and not before, the main body of economic analysis; and though it may have its beginning within the proper domain of economics, it cannot find its conclusion there, but must extend far beyond.

Thus, economics was to be only a piece of a far larger project, a crucial piece, to be sure, but only a piece. We start with what is within our grasp, with what is simple, and when that is fully mastered, we move on. This was Marshall's vision. But as is so often the case in science, that vision got blurred. Economic science is still trying to master the simple, with formulations that grow increasingly complex. That this was only going to be a piece has been forgotten.

We can sketch the principles of highly scientific, quantitative, neoclassical economic science as they emerged in the late nineteenth century. They are wholly continuous with the views of Adam Smith. They begin with the view that economic science is morally neutral. Like all science, economics is concerned with the facts, with discovering and describing the way things *are*, and with predicting the way things *will be* if one or another practice is followed. It is not concerned with oughts, with the way things *should* be. Economics is out to reveal the laws of economic human nature. It is out to describe how economies, markets, firms, and individuals work. A successful economic science will eventually be able to tell people what to do, as individuals, firms, or economies, to achieve their goals. It will not be able to tell them, nor does it aspire to tell them, what their goals should be. It is a science of means, means to satisfy whatever wants people happen to have.

This moral neutrality of economics requires elaboration. It is not simply the result of trying to be scientific. Rather, it reflects some significant assumptions about human nature and human desires. How could economics tell people what they ought to do? Well, as a science of means, it could tell people what they ought to do if it knew what their goals were. If it knew what people needed, what was valuable to people, it could figure out the best means to meet those needs, or achieve those values. It could tell them that X is the best means to Y, and that they therefore ought to do X, because they need Y, or because Y is valuable. But economics insists that it has no theory of need or value. Indeed, there is no theory to be had. Different people will value or want different things. Some of the things that people say they "need" will be culturally defined. Everyone in this culture may need a Michael Jackson poster, but that will not be true in all cultures. Some "needs" will be determined by what other people have, as for example, people in our own culture seem increasingly to need personal computers or videocassette recorders. In everyday discourse, people make distinctions between what they *need* and what they merely *want,* but economics tells us that such distinctions will not stand up under close scrutiny. As a result, economists do not even distinguish needs and wants. All human desires are treated as wants. Furthermore, these various wants that people have cannot be ranked. Neither economists nor anyone else can say that one want is better or more important than other. The value of the different things that people want will vary from one person to another. Indeed, the best way to determine the value of various things is by seeing what people will pay for them in the market. The presumption is that the individual is the best judge of his or her own welfare and will act in the market so as to promote that welfare. The market then is not simply the most efficient means for arranging exchange between people. It is the means for determining economic value. It may be the only yardstick for determining value available.

The idea that each individual is the best judge of his own welfare has profound implications. It implies that there is no way to make interpersonal comparisons of welfare. There is simply no way of determining whether the things I want mean as much to me as the things you want mean to you. If I want to drive a huge, gas-guzzling car to work, and you want less traffic, less pollution, and plentiful fuel, there is no way of deciding whether the benefits to me are greater than the costs to you if you give up your wants while I satisfy mine. And if the welfare of society as a whole is just the sum of the welfare of its individual members, what this incommensurability of individual wants implies is that there is no way to say what is good for society as a whole. Unless some social practice can be determined to increase the welfare of some people without decreasing the welfare of any, one cannot be sure that the social practice constitutes a net improvement in the welfare of society. In the extreme, if some change in the tax law eliminated malnutrition in millions, at the cost of a few elegant dinners for thousands, one could not be sure that this tax law increased the welfare of society.

Now that we have a picture of economics as value neutral science, we can ask what it is a value neutral science of. What are these creatures who go about attempting to satisfy their unlimited wants? Smith called them economic men. Modern economics has replaced Smith's crude notion of economic man in pursuit of want satisfaction with a refined picture of *rational economic man*. Rational economic men don't merely pursue the satisfaction of wants. They act always so as to *maximize* the satisfaction of wants. More specifically, rational economic agents can be said to possess the following characteristics:

1. *Rational economic agents can always express preferences between commodities, or bundles of commodities.* When asked whether he prefers commodity A to commodity B, a rational economic agent can always give an answer, either yes, no, or indifferent. Saying that he is indifferent between A and B is not saying that he can't express a preference; it is saying that he doesn't care, that he would just as soon have one as the other. Though economic rationality requires that people be able to state a preference between A and B, it does not require that they be able to say *how much* they prefer one to the other. That is, people are able to give what are called *ordinal* rankings to various commodities, on the basis of how much satisfaction, or utility, they provide. This ordinality of preferences is a lot like the rankings made of athletic teams. Everyone may be confident that Team A is better than Team B, which in turn is better than Team C, but be completely at a loss as to how much better one team is than another.

People's ability to express preferences need not extend to all aspects of their lives. For example, someone may not be able to say whether he prefers good looks, intelligence, or wealth in a prospective mate. However, within the economic sphere, within the domain of goods and services, as rational economic agents, people must be able to express preferences. We will see later that this qualification can sometimes pose serious problems for an economic analysis, as for example when people are asked to choose between things that have both economic and noneconomic aspects. Consider having to choose between two jobs, one of which offers good pay and benefits and security for boring work with uninteresting colleagues, while the other offers low pay and benefits and little security for demanding work with interesting colleagues. We may find ourselves at a loss to express preferences here, since it requires an assessment and weighting of economic aspects of the job (salary and benefits) as well as noneconomic ones (potential satisfaction and social stimulation). But leaving aside this complication for now, rational economic agents should be able to express ordinal preferences among various combinations of commodoties.

2. *Rational economic agents should prefer more of a desirable commodity to less.* If one apple is good, two are better, and three are better

still. This is sometimes called the "monotonicity" assumption, and it seems obvious. If some quantity of a thing is good, more of that thing is better, no matter what the thing.

3. *Rational economic agents should prefer a lower price to a higher one.* This too seems obvious. A dozen apples for two dollars will be preferred to the same dozen apples for three dollars.

4. *The rational economic agent always chooses the most preferred commodity or combination of commodities from those that are available.* This tenet of economic rationality is important because it makes possible the diagnosis of preferences. How, after all, does one know what people want? One way to know is to ask them. People could be given an array of hypothetical choices, and from their responses a picture of their preference rankings could be constructed. But hypothetical choices are just that—hypothetical. It is a big step to assume that what people choose hypothetically is what they will choose when they actually have their money on the line. An alternative route to diagnosing preferences is to watch what people actually *do* choose. Their choice behavior, coupled with the assumption that choices reveal underlying preferences, make it possible to infer what their preference rankings must be. Furthermore, economists are not generally interested in the economic decision making of individual people; they are interested in the economic activity of large groups—even whole societies. What a monumental task it would be to determine the preference rankings of a society by giving each of its members an array of hypothetical choices. By assuming that choices reflect preferences, this elaborate polling becomes unnecessary. One can diagnose societal preference structures simply by keeping track of what people actually buy in the marketplace.

5. *The preferences of rational economic agents are relatively stable over time.* If the pursuit of self-interest is to bring some semblance of order to an otherwise chaotic world, this tenet of economic rationality must be true. Individuals can't prefer apples to oranges one day and oranges to apples the next. The possibility of rapid shifts in preference would make people's behavior in the marketplace appear unpredictable, even incoherent. There would be no way for a potential trader in the market to guess at what people might be willing to pay for the commodity he is offering. There would be no way for firms to determine how much of a given commodity to produce. Indeed, unless preferences

were stable, the very notion of rationality would lose all meaning. What would it mean to say that people always choose the most preferred alternative, if that alternative could change every day, or every ten minutes? In the absence of stability, any pattern of choice whatsoever would have to be viewed as rational. Even blatant inconsistencies could be interpreted as reflecting changes in preference.

Saying that people's preferences are relatively stable is not saying much. How stable is "relatively stable"? Relative to what? Why introduce the qualifier *relatively* at all? Why not simply argue for absolute stability of preference as a feature of economic rationality? The reason is that no one has absolutely stable preferences. People change. They acquire tastes for some things and lose tastes for others. So any framework that did not allow for changes in preference would be woefully impoverished. We are stuck, then, with the less than precise idea that while rational agents may change their preferences from year to year, or even from month to month, they do not change them from day to day, or hour to hour.

6. *Rational economic agents have preferences that are transitive.* If someone prefers apples to oranges, and prefers oranges to bananas, then if she is rational, she'd better prefer apples to bananas. More generally, transitivity of preference means that if A is preferred to B, and B is preferred to C, then A is preferred to C. Transitivity of preference as a feature of economic rationality confers a great deal of predictive power on a theory of economic behavior. We don't actually need to observe people making choices among all possible sets of commodities. We can observe some choices and use the assumption of transitivity to make inferences about what other choices would have been had the opportunities for making them arisen. Transitivity is a feature of many of the relations people use to order things in the world. The relation "bigger than" is transitive; the relation "older than" is transitive; the relation "richer than" is transitive. But not all relations are transitive. For example, "has authority over" is not a transitive relation. Neither is "is a friend of," or "is in love with." The claim of the economist is that commodity preference is a transitive relation.

7. *The more that rational economic agents have of a commodity, the less that further increases in that commodity contribute to their satisfaction.* This is known as the principle of diminishing marginal utility. A loaf of bread is worth a great deal to a starving man. The second

loaf is also worth a lot, though less than the first. The third loaf is worth still less, the tenth, still less, and the hundredth, perhaps little or nothing. The more a person has of something, the less interested he is in having more of it. This principle of economic rationality is of central importance. Think for a moment about two people deciding whether to trade in the marketplace. One of them has eggs to trade, while the other has cheese. Will they be able to strike a bargain, trading so much cheese for so many eggs? Since people are free to do whatever they want in the market, no bargain will be struck unless both parties feel they are gaining from the exchange. No rational person would make a trade that left him worse off. Indeed, no one would make a trade if it didn't leave him better off. There is no point spending time and energy in the business of exchange if the best a person can do is end up breaking even. But how can both parties be made better off from a transaction? If one person is going to profit, must it not be at the expense of the other? If a dozen eggs are worth a pound of cheese, and both parties know that, how can the trade end as anything but a stalemate?

The principle of diminishing marginal utility provides the answer to these questions. It provides the grease that lubricates market transactions. For in fact, a dozen eggs are worth more to the cheese seller than they are to the egg seller. For the cheese seller, they will be his first dozen; for the egg seller, they are just one dozen among hundreds. Conversely, a pound of cheese is worth more to the egg seller than it is to the cheese seller. They can make a fair exchange, and both wind up better off. *This* dozen eggs will provide more utility or satisfaction to the cheese seller than they will to the egg seller, and *this* pound of cheese will provide more satisfaction to the egg seller than it will to the cheese seller. So both traders make out well from the transaction.

The principle of diminishing marginal utility applies to all commodities in the marketplace. It is a force that drives toward equity in the distribution of goods and services because it makes people willing to give up some of what they have a lot of for relatively little in return. Through the market, the principle of diminishing marginal utility works to allocate resources where they will do the most good, that is, provide the greatest satisfaction. The one significant exception to this principle may be money. Since money can in essence become any commodity, its utility may not diminish, or may diminish much more slowly, than the utility of other commodities, though even in the case of money, people with plenty of it seem willing to trade large quan-

tities for goods and services that others, with less money, wouldn't
dream of purchasing.

8. *A rational economic agent acts in the marketplace on the basis of
complete information.* This means that rational agents know everything about what is available, for what price, now, and about what
will be available, for what price, in the future. This notion that people in the market have perfect information is, of course, an abstraction, an idealization. In actual fact, no one has perfect information.
This abstraction views market behavior as a little like choosing a dish
in a Chinese restaurant. All the possibilities are arrayed on the menu,
the buyer knows exactly what everything is and how much it costs,
and his rational choice will be that item that provides the greatest
satisfaction for the lowest price. This abstraction is no doubt unrealistic, but it is nevertheless important to economic theorizing. Unless
one knows what a person in the marketplace knows, or one assumes
that the person knows everything, it is very difficult to predict how
the person will act. This is true even if one grants the other characteristics of rationality already mentioned. It is hard to assume that a
person chooses what he most prefers unless we can be sure that he
knows what the possibilities are. Someone may choose a low-fare,
three-stop flight from Philadelphia to Los Angeles because she doesn't
know that a nonstop flight at equivalent fare is available from an
obscure airline. More generally, it is very difficult to determine whether
seemingly peculiar market choices are the result of deviations from
rationality by fully knowledgeable people or the result of full-blown
rationality by ill-informed people. Assuming that people are rational
doesn't do much work for the economist under these conditions of
uncertainty.

A way to make this abstract idea that people act with perfect information more concrete and more realistic is to treat information itself
as a commodity and to ask how much time and effort people are willing to put into gathering it. That is, information gathering has costs,
and these costs may make it so that the rational individual compromises on just how much information he will seek before acting. The
economists task then becomes one of figuring out some way of assessing the costs of information gathering and comparing them with the
benefits of having that information available. If this approach does
not seem more realistic, think about the costs involved in determining
what the best possible personal computer to serve an individual's needs
is, or the best automobile, or the best stereo system. A person could

spend his whole life on a quest for perfect information—about product quality or price—and end up with no time left to do anything with that information. So information costs are indeed real, and some economists worry about them. But in the absence of information about information, the economist assumes that rational agents have all relevant information available.

9. *Rational economic agents use their rationality to make choices in the market that maximize the service of their preferences.* We come, finally, to the nub of economic rationality. What is it all for? What is the goal that economic rationality serves? The answer to this question has changed over the years. For a time, it was thought that economic activity was exclusively in the service of pleasure, that people were essentially hedonists whose interest was in feeling good and whose rational activity served that interest. Thus, economist Stanley Jevons could say, in 1871: "To satisfy our wants to the utmost with the least effort—to procure the greatest amount of what is desirable at the expense of the least that is undesirable—in other words, to *maximize* pleasure, is the problem of economics."

Furthermore, Jevons argued, the principles of rational economic man in pursuit of pleasure are "so simple in their foundation that they would apply more or less completely to all human beings of whom we have any knowledge." Over time, however, economists came to realize that not all activities seemed to be serving the quest for pleasure. Some economic choices were made because the results were *useful,* and pleasure really seemed beside the point. Someone might choose to live close to where she works, not because a short commute is more pleasurable than a long one but because it is more useful, or convenient. Introducing "pleasure" seemed gratuitous. This insight led to a more general formulation that economic activity serves the maximization of *utility.* Sometimes, but not always, the utility-maximizing choice would bring pleasure, but it was utility, not pleasure, that economic agents, at bottom, were after. The notion that people are utility maximizers seems more reasonable than the notion that they are pleasure maximizers, but the concept of utility is not without its problems. For one thing, how does one measure utility? An objective measure of utility would seem to require some idea of what is valuable to people. But we have already seen that economists reject the idea that there is a general theory of human value. Different people will value different things, and all there is to go on in determining what an individual values is what he chooses. Given this view, the notion of utility loses meaning. If people are rational, that is, if they are choosing what they prefer, then by definition they are maximizing utility. We *see* people maximizing preference, and we *infer* that they are maximizing utility. So why not dispense with the inference altogether, and view the rational economic agent as a maximizer of preference? This is the conception adopted by virtually all modern economists.

This, then, is the rich and detailed picture of economic rationality that modern economics presents. People enter the marketplace prepared to apply these rational tools to the service of their interests, and their interests are to maximize their preferences. People consider everything that is available now (full information); they consider what might be available if they defer consumption until later; they consider the costs and benefits of various alternative actions, perform the necessary calculations, and come finally to a decision about which allocation of resources will maximize their preferences and bring the greatest satisfaction. They are never satisfied, that is, they always possess some want that they would like to have met. As a result, they never really leave the marketplace.

Economic Rationality and Labor

Economic rationality extends not just to the consumption of goods and services but also to decisions about work. Work is regarded by economists as a cost, a *disutility*. It is just the means by which people accumulate the resources that can then be devoted to satisfying preferences. As a result, decisions about work will be largely determined by the material rewards it makes possible. Salary, benefits, security, chances for advancement, and the like will be weighed against commuting costs and working hours to determine which job yields the greatest net benefits. Sometimes, of course, choice of work will be influenced by other factors, factors that are not obviously economic. A man may choose a job that allows him to work among friends because the social interaction it provides will make the job less onerous. He may choose a job in which he doesn't have to work so hard so that he has energy left to enjoy the fruits of his labors. He may choose a job with flexible working hours, perhaps to accommodate other interests he has. He may choose a job that promotes human welfare, out of some sense of responsibility to society that he feels. He may choose a job because the work is challenging and varied, because that will make the (psychological) costs of working lower. He may choose a job because of the status or prestige it confers. Any or all of these considerations may influence a person's choice of work, along with the material consequences of what he does.

Economists have two different ways of dealing with these kinds of nonmarket influences on work. One is to acknowledge them as noneconomic, to acknowledge that work is not a pure disutility, and that various noneconomic considerations can affect how much intrinsic satisfaction work brings, and thus what work people choose. Considerations like these would then be regarded by economists as *externalities,* factors external to economic rationality that nevertheless impinge on economic decision making, so that a purely economic analysis of work choice would be incomplete. This might be seen as conceding a lot about work to noneconomic factors, but the economists' view is that, generally, these factors are of only marginal importance. Either they are very important, but only to a handful of people, or they are only important when all economic considerations are equal. That is, the majority of people may worry about prestige, social stimulation, flexibility of hours, and the like only when choosing between jobs that provide essentially equivalent material benefits. In support of this view that noneconomic factors are of only marginal importance in one's choice of work, the economist can point to the sorts of issues that are almost invariably on the table when management and labor unions negotiate contracts. With only the rarest of exceptions, it is material costs and benefits that are subject to negotiation. Furthermore, standard management practices work to keep these noneconomic factors at only marginal significance. It is a common policy to shift supervisors around from place to place to prevent noneconomic social relations from developing between boss and worker, or between workers. Management is interested in ensuring that the coin of the realm in the workplace is strictly material, in part because material benefits can be strictly controlled, and with them, the behavior of the workers can be predicted and controlled as well.

This conception of work as basically a pure means to material ends has been a part of economic thought for a long time. Adam Smith had this to say:

> It is in the inherent interest of every man to live as much at his ease as he can; and if his emoluments are to be precisely the same whether he does or does not perform some very laborious duty, to perform it in as careless and slovenly a manner that authority will permit.

On this view, the only "cost" of unemployment is the welfare check that the unemployed worker receives. Not working, in itself, is a pure plus. Indeed, one of the things one could imagine rational economic agents making choices about is how much time to spend working and how much time to spend at leisure. One could imagine an infinite number of "bundles," each containing different proportions of labor time (which translates into money) and leisure time. Everyone's optimum, it is assumed, is 100 percent leisure and 0 percent labor. The problem is that only the independently wealthy can make this choice. Everyone else chooses as much leisure as can possibly be afforded. Suppose that someone chooses a bundle that is half labor and half leisure, eight hours per day of each (the rest of the day is sleep, which won't count). This gives her the optimal mix of commodities (purchased with salary) and time to enjoy them. Now further suppose that she gets a raise, from say ten dollars an hour to fifteen dollars. What should happen to her labor leisure trade-off now? At first blush, it might seem that since she can now earn the money she wants in less time than before, she will opt to work less and play more. Indeed, if she strikes a compromise, she can earn a little more money than she did before and have more time in which to enjoy it. On the other hand, her pay increase has increased the price of leisure, in wages passed up when she enjoys it. Each hour of leisure now costs fifteen dollars instead of ten. Economic rationality dictates that, all other things being equal, as prices go up, consumption will go down. This implies that her pay increase will make her work more. So which is it? Do people work more or less when their pay increases? The answer is that there is no general answer. It could go either way, depending on which of the two opposed effects is stronger. For the present point, though, it doesn't really matter which way it goes. Either way, the underlying presumpton is that work is a pure disutility.

There is a second thing that an economist might say about so-called noneconomic influences on work choice that seem to make work not a pure means to an end. He might say that they are *not* noneconomic. Each of these factors has its price. If prestige is important to someone, he should be willing to pay for it, in the form of decreased wages and benefits. The same is true of flexible working hours, challenging job demands, social stimulation, and the like. Each of these features now becomes part of the market. Employers "sell" attractive job conditions; employees "buy" them. The commodity bundle containing a twenty thousand dollar salary and flexible hours may simply be pre-

ferred to the commodity bundle containing a twenty-five thousand
dollar salary and rigid hours. In principle, one could even quantify
these nonmaterial influences on work choice by determining just how
much salary workers would be willing to pay for each of them.

Now this move is an extremely significant one. Once the economist
argues that such things as social contact, intellectual challenge, and
the like in the workplace can be given a market price, it is tempting
to ask why their economic nature should be restricted to the workplace. Why not inquire about what people are willing to pay for social
contact, intellectual challenge, and prestige in general? Recall that,
originally, economists viewed economic nature as just a part of human
nature—a significant and autonomous part to be sure, but just a part.
Well, if such things as social contact can be priced, then there seems
to be no limit to how far economic nature extends. Human nature
may be economic nature through and through. Everything people do
has its price. Every activity has its market. Such activities as love and
marriage, religious pursuit, children, and patriotic loyalty, can be
analyzed in terms of their costs and benefits. They can be treated as
no different, in principle, from video recorders and personal computers. All aspects of a person's life can be measured by the same yardstick. The commodity bundles a person chooses among can involve
choices like the following: high prestige, high salary, hard work, cut-
throat competition; fancy house, expensive car, good clothes, restaurants, and vacations; no spouse, no children; versus, high prestige,
moderate salary, moderate work; plain house, car, and clothes; spouse,
and family. Where previously this could be thought of as a choice
between playing economic man to the hilt and not doing so, now it
can be seen as simply a choice between two sets of commodities.
Everyone plays economic man to the hilt; people just have different
preference hierarchies.

Not every economist embraces this extension of economic rationality to all aspects of life. Many are content to keep their formulations
restricted to the everyday domain of goods and services. But there is
certainly a move afoot to spread economic analysis to everything. The
spirit of this development is well captured in a discussion of the economics of marriage, by economist Gary Becker:

> By assumption, each marital "strategy" produces a known amount of
> full wealth (i.e., money wealth and value of nonmarket time), and the
> opportunity set equals the set of full wealths produced by all conceiv-
> able marital strategies. The individual ranks all strategies by their full

wealth and chooses the highest. Even with certainty, a strategy with marriage then dissolution, and eventually remarriage might be preferred to all other strategies and would be anticipated at the time of first marriage. Dissolution would be a response perhaps to the growing up of children, or to the diminishing marginal utility from living with the same person, and would be a fully anticipated part of the variation in marital status over the life cycle.

Thus, according to Becker, whether and when to marry, whether and when to divorce, and whether and when to remarry, are all part of cost–benefit calculations that are designed to maximize our preferences, or what he calls full wealth. This constitutes a major step in the making of "unidimensional man," a person who in every aspect is engaged in and guided by rational, economic consideration.

Economic Rationality and the Market

Adam Smith's insight was that if a bunch of self-interested, rational economic individuals were thrown together in the free market, the results of their exchanges would be good both for them and for society as a whole. Indeed, Smith argued, as many have argued since, not only would the collective results of individual free-market activity be good, but they would be the best possible results. The free market maximizes the preferences not just of individuals but of whole societies. Now that we have a detailed understanding of what it means to be a rational economic agent, let us examine under what conditions the market is free and allows economic agents to operate effectively.

1. The free market consists of a large number of small and anonymous firms. Firms must be small so that no one of them can dominate the market, driving others out. Market choices are governed by price, and prices are governed by competition. It is the competition among firms for the consumer's dollar that drives them to increase their efficiency and productivity. If one firm were so large that its activity completely altered the shape of the market, or affected the availability of raw production materials to all other firms, competition would be impaired, and with it efficiency and productivity.

2. There must be little product differentiation. If every brand of chocolate chip cookies is significantly different from every other, then the cookie makers are not really in competition with each other. A choice between cookie A and cookie B would be like a choice between cookies and rolls. Real competition requires essentially equivalent

products so that consumers will be motivated in the market to find their best deal.

3. Resources must be mobile. Producers must be able to shift what they produce in response to changes in market demand. A manufacturer should be able to capitalize on the hula hoop craze by changing his plant from a producer of radios to a producer of hula hoops, and he must be able to do this quickly and at relatively little cost. If resources are not very mobile, then the market's response to changes in demand will be sluggish and costly.

4. Both producers and consumers must act with perfect knowledge, both of the present and of the future. Mistakes in estimating current or future demand will result in market inefficiencies. Consumers will get more hula hoops than they want, and fewer radios. Mistakes in keeping track of what is available will lead consumers to pay more than they have to, thus allowing producers to get away with production inefficiencies.

5. Goods available in the market must be *rival,* that is, if one person consumes a good, others cannot also. Goods must also obey the principle of *exclusion,* that is, only the people who pay for a good derive its benefits. For goods that do not possess these two characteristics, the market mechanism will not work effectively. In general, the market will not work if people can get what they don't pay for, if they can be "free riders" on the backs of others. The relation between supply, demand, and price will be thrown out of kilter if some people will be getting for nothing what others must purchase.

6. Participants in the market must have fairly equal resources. Think of a poker game. All other things equal, the best hand usually wins. And over the long haul, the best players beat the worst ones. But there is something that can beat even good hands held by good players. That is rich players. Rich players can simply force good ones out of the game by betting in amounts that they can't match. The same is true in the market. Rich people can have sectors of the market all to themselves by simply bidding prices up so high that most others are excluded from participating. People with a substantially disproportionate share of resources can always approriate whatever piece of the market they like for themselves, unless some extra-market force (law, government, ethics) intervenes to prevent it. The rich always have the opportunity to "corner the market."

7. Firms in the marketplace must be out to maximize profit, just as individuals are out to maximize preferences. Supply, demand, price, and competition will lead to efficient production of what people want

only if all the players in the game are playing to win. And what it means to "win" the economic game is to make as much profit as possible—to sell as dear as possible and buy as cheap as possible.

When a market possesses these characteristics, several things will be achieved. First, the market for all goods will clear. That is, prices will move to a level where supply equals demand. Excess supply will lower prices; excess demand will raise them. At equilibrium, there will be exactly one hula hoop for everyone who wants one. This market clearing applies to the market for labor as well. There will always be work for everyone who wants it. Supply and demand will determine the price of labor, the going wage. What this implies is that what unemployment there is under free-market conditions is voluntary. The unemployed worker could always get a job by offering his services for less than the going rate. And what this in turn implies is that deep down, there is no such thing as power in the workplace. There are no bosses and workers. It could as well be said that the worker hires a boss as that the boss hires a worker. This is because with the conditions of work determined in the market, the boss has no more control over wage levels than the worker does. A boss who pays less than the market wage will find his workers going elsewhere; a boss who pays more than the market wage will find unemployed workers trying to bid his wage rate down. At equilibrium, everyone will pay the market wage, and everyone will earn the market wage, as a result not of anyone's controlling hands but of the market's invisible ones.

Economic Man, the Market, and Social Welfare

It is a guiding assumption of economics that economic rationality is a characteristic of human nature. Rational economic agents are born, not made. Human motivation to engage in cost–benefit calculation in the service of maximizing preferences is an unalterable part of us, as much a part of what it means to be human as the fact that people speak, have opposable thumbs, and walk on two legs. The pursuit of self-interest requires no more justification than the fact that planets revolve around the sun.

The same cannot, however, be said of the market. The free market is a human creation. Societies could arrange themselves differently, as many obviously do. Societies could encourage trade of some things, restrict trade of others, and prohibit trade of still others. Even in America, the hallmark of the free market, there are plenty of restraints on market activity. Minimum wage laws, welfare benefits, subsidized health care, public education, and the like are all restraints on the market. So is legislation that breaks up some monopolies and allows and regulates others. Finally, American society simply outlaws trade of some commodities. These commodities are called *inalienable,* meaning that people can't alienate them from themselves. They can't part with them or sell them. People can't sell their children; they can't sell their bodies for sexual service; they can't sell themselves into slavery. What this all means is that marketlike arrangements are subject to human discretion. Each society must decide for itself just how free the market will be, and in what domains it will be allowed to operate. Each society must decide for itself whether to allow the distribution of resources among its citizens to be determined by the market or to impose restraints on distribution. Philosopher John Stuart Mill said this over a century ago in his book on political economy:

> Distribution of wealth, therefore, depends upon the laws and customs of society. The rules by which it is determined are what the opinions and feelings of the ruling portion of the community make them, and are very different in different ages and countries, and might be still more different if mankind so choose.

Much of the debate that currently goes on in American society about economic matters is concerned precisely with the question of distribution of wealth. At present, the market is not allowed free rein in determining who gets what. Taxation is used to take resources disproportionately from the wealthy, and various welfare programs are used to redistribute them to the poor. The state offers free public education to everyone. Since education is financed by taxation, this is another way of redistributing wealth from the rich to the poor. Those who are identified as economically conservative argue that there is too much restraint on the free market. Taxation should be lower. People should be able to keep what they earn and buy what they want. The government should stay out of business. Now what is the force behind these various "shoulds"? Why "should" the government do this rather than that? How can an unfettered free market, or a restrained one, be defended?

There are actually two quite distinct kinds of arguments that are made to back up these various assertions about what the government should or should not do. One of them is moral and has really nothing in particular to do with economics. It is based upon ideas about what people in a free and open society are entitled to, what individual rights are. The second line of argument is not moral; it is technical. It claims that the free market is the most efficient and effective possible arrangement for determining the allocation of scarce resources. Any other kind of arrangement will bring social costs that more than outweigh social benefits. This line of argument is squarely within the province of economics. Just as economic rationality is the way to maximize individual satisfaction, the free market is the way to maximize social satisfaction. Let us examine these two lines of defense of the free market, beginning first with the technical, economic defense based on efficiency.

The branch of economics that is concerned with assessing the efficiency and effectiveness of various economic schemes is known as welfare economics. It asks, in essence, about what sort of market arrangements will maximize social welfare. Answering this question is extremely difficult. Indeed, even asking it in a form that might permit an answer to emerge is extremely difficult.

The difficulty comes in measuring social welfare. If one starts by assuming that the collective welfare of a society is just the sum of the individual welfares of its members, one has the task of measuring individual welfares. But as we have already seen, this is no simple matter. There is no formula to determine how much welfare a particular commodity will confer on its owner. Different people want different things, or if they want the same thing, they want it in different amounts, or to different degrees. Individual wants, remember, are incommensurable, incomparable. And individual utilities can only be ranked ordinally. It is possible to judge whether someone likes A better than B, but now how much better. Furthermore, sometimes it is not clear that the plusses and minuses of different alternatives can even be compared on the same scale, even within an individual, let alone between them. Suppose society has a pile of money to spend, and the question is whether to spend it to improve education or to spend it to improve people's mental health. How can the costs of doing poorly in school be compared with the costs of being pathologically depressed? Or suppose a factory owner has to decide whether or not to introduce a new, high-powered production method into her factory, one that will increase people's wages, but keep them under a great deal of pressure. How does she weigh the gains in productivity and pay against the costs in happiness and morale? How then, in short, is it possible to get a fix on individual welfares, let alone collective welfares?

Since one of the problems here is comparing one person's costs against another person's benefits, a solution to the problem is to look for situations where there are benefits without costs. Economist Vilfredo Pareto has formulated principles to guide the assessment of welfare in such situations. According to Pareto, a social optimum exists when no one in society can be made better off without someone being made worse off. This simple principle can be used to determine whether one or another proposed change in economic policy will provide an improvement in social welfare. It is certain that the policy will be an improvement if at least one person will be made better off with no one being made worse off. This has come to be known as the *Pareto criterion*.

The Pareto criterion can be used to defend the free market as the most efficient economic arrangement. The argument runs something like this: By definition, if people are allowed to trade freely, they will only agree to a trade if it will make them better off. At worst, one of the parties in the trade will end up even while the other benefits. So any freely agreed upon trade will be Pareto efficient, in that it satisfies the Pareto criterion that at least one person will be better off, while no one is worse off. When will trading stop? When neither party stands to gain any more from trading, that is, when the Pareto criterion is no longer satisfied. No other economic arrangement, no arrangement that restrains free trade, can be assured of meeting the Pareto criterion. In any arrangement involving restraint, at least some situations will arise in which someone ends up worse off as a result of exchange. Consider, for example, minimum wage laws. Suppose the minimum wage is set at three dollars an hour. Without it, three teenagers might agree to trade their labor for a wage of two dollars an hour. With it, the employer can only afford to hire two of them. These two end up better off, but the third ends up worse off. Or consider the regulation of prices, for say, electricity. Assuming that regulation keeps prices down, users of the service benefit from regulation. But the company, and those who own stock in it, lose. If prices were market determined, people would buy as much electricity as they wanted at a given price, and the company and its stockholders would make greater profits. So the market and only the market is Pareto efficient, because the market and only the market essentially has Pareto efficiency built into it.

What of the moral defense of the market? The moral defense focuses on the value of freedom of choice and on people's rights and entitlements. It is at least as old as philosopher John Locke, who wrote three centuries ago. More recently, it has been articulated by Robert

Nozick, in his book, *Anarchy, State, and Utopia.* The idea, roughly, is this: people are entitled to keep the fruits of their labors; they have a right to property. If a market system coerces no one, that is, if people are at all times free to accept an exchange or reject it, then the *process* of exchange is fair and just. This does not mean that the *results* of exchanges will necessarily be fortunate or desirable. One party to the exchange may be shrewder or luckier than another. People who are consistently shrewd will amass great wealth, while people who are consistently foolish will be poor. Everyone may agree that vast differences in wealth within a society are unfortunate. However, the argument goes, they are not unjust. Furthermore, any effort to correct unfortunate results with forced redistribution (as through taxation) will be unjust, for it will involve forcibly taking from people what is rightfully theirs, what they have earned through hard work or cleverness. In a word, if one is after equality of opportunity, the free-market system is a fair and just one—perhaps the only fair and just one. But if one is after equality of outcomes, fairness, justice, and the free market will have to be sacrificed.

There are, of course, other moral stories that can be told that lead to quite different judgments about the free market and justice. One of them, told by John Rawls in his book *A Theory of Justice,* argues that a just system of distribution is one that people would choose if their own position in that distribution were to be determined at random. That is, if a person didn't know whether she was the daughter of a beggar or of a millionaire and she could opt for any pattern of resource distribution in society at all, what would she choose? Rawls suggests that under these conditions, people would look for equality of outcome rather than equality of opportunity. He further suggests that choice from this "original position" of ignorance about one's own status in society is the proper test of what is just.

This is not the place for deciding between the moral defense of the free market and various alternatives to it. The important points to realize here are these: first, it is possible to offer a defense of the free market on moral grounds; and second, such a defense is not open to economics, at least not if economics is to be morally neutral science. The economist's defense of the market must be that it works better, given the rational, economic nature of human beings, than anything else, and not that it is just. We will examine later just how plausible this "scientific" defense is, as we examine the plausibility of rational, economic, human nature.

Evolutionary Biology and Human Nature

The evolution of society fits the Darwinian paradigm in its most individualistic form. . . . The economy of nature is competitive from beginning to end. . . . No hint of genuine charity ameliorates our vision of society, once sentimentalism has been laid aside. What passes for cooperation turns out to be a mixture of opportunism and exploitation. . . . Scratch an altruist and watch a hypocrite bleed.

MICHAEL GHISELIN

I never yet touched a fig leaf that didn't turn into a price tag.

SAUL BELLOW

There are many species of birds that try to get off easy in the chore of childrearing. They deposit their eggs in the nests of other species of birds and let those other species do the work. The problem, of course, is that the parents who are overseeing these other nests are not especially pleased to have extra, alien mouths to feed. So they remove these foreign eggs from the nest whenever they can recognize them as foreign. This reluctance on the part of the "foster parents" puts pressure on the egg abandoners to produce eggs that look just like the eggs of the foster parents, so that recognition and rejection of foreign eggs is impossible. This, in turn, puts pressure on the foster parents to sharpen their ability to spot aliens.

An exception to this pattern occurs in the case of a species of cowbird that lays its eggs in blackbird nests. These blackbirds are constantly endangered by botflies, parasites that destroy their young. The blackbirds typically solve this problem by nesting near wasps or bees, insects that happen to repel botflies, thus protecting the blackbirds. Without this protection provided by wasps and bees, the blackbirds are in big trouble—*unless,* that is, their nests are invaded by cowbirds. For cowbird young will remove botflies from themselves and from their foster siblings. In this case, taking care of cowbird young actually does the blackbird some good.

However, this does not lead the blackbird to accept cowbird eggs indiscriminately. Blackbirds whose nests are near the nests of wasps and bees will destroy cowbird eggs. Only blackbirds whose nests do not neighbor the nests of wasps and bees will take the cowbirds in. So there are two sorts of blackbirds: the ones who nest near wasps and bees and reject cowbirds, and the ones who don't nest near wasps and bees and accept cowbirds.

And to go with these types of blackbirds, there are matching types of cowbirds. Some cowbirds deposit their eggs only in blackbird nests that are away from wasps and bees. These cowbird eggs don't look anything like blackbird eggs. But, of course, they don't have to, since the blackbirds are quite eager to have the cowbirds anyway. The other cowbirds deposit their eggs in the nests of blackbirds that are near wasps and bees. These cowbird eggs are almost perfect mimics of blackbird eggs. They'd better be, or else the blackbirds will dispose of them.

All of this is extraordinarily sensible. How clever of the cowbirds to get rid of botflies. How shrewd of them to go to the trouble of producing eggs that look like blackbird eggs only when they have to. How well conceived that the blackbirds tolerate cowbirds when, and only when, it does the blackbirds some good. How could these little birds be so smart? How do they do it?

Until the nineteenth century, there were two sorts of answers to these questions. As naturalists explored the world of living things, they encountered one example after another of animals that seemed to have found the perfect solution to the problems of survival they faced. Their most common answer to the question "How?" was an appeal to the hand of God. God had designed all living things and the environments in which they lived. Why would he bother designing them without giving them the equipment they needed to get along? Why put fish in the water without giving them gills? Why put birds in the air without giving them wings? The exquisite fit between organism and environment was no accident. It occurred because both were a part of the same master plan, the same grand design, implemented by an all-knowing and all-powerful God.

The alternative to this argument from design, as it is sometimes called, was to imbue animals with the same kind of intellectual capacity that people had. The cowbird and the blackbird "figured it out." They were players in a chess game, and they went about the game trying to anticipate each other's moves and to develop winning strategies based upon their anticipations. This kind of answer to the question "How?" *anthropomorphized* animal behavior, assuming that behavior that *looked* as intelligent as human behavior *was* as intelligent as human behavior. People do this sort of thing all the time, ascribing a variety of humanlike desires, intentions, expectations, plans, and emotions to pet dogs and cats.

Until the nineteenth century, these were the two alternatives; design from the hand of God, or the product of rational intelligence. Then came Charles Darwin. Darwin's theory of evolution by natural selection answered the question "How?" without appeal either to God or to godlike intelligence. It spawned the discipline of evolutionary biology and its accompanying vision that the pursuit of self-interest was a universal, natural law. This vision, in its turn, provides critical support for the economist's conception of rational economic man. Let us then examine what evolutionary biology proposes to tell us about human nature.

Darwin

Darwin's theory contained two crucial components. The first component was *variation:* individual members of a species differed from one another in structure and in behavior. The differences weren't huge, but they were there. Some members of a species were a little bigger than others, or a little stronger, or a little faster, or a little less selective about what they ate, or even a little smarter. The second component was *selection*. Not all the members of a species were equally well equipped for survival and reproduction. The stronger, faster, smarter ones had an advantage over their fellow species members. They would play this advantage out by having more offspring, who would also be strong, fast, and smart. Over many, many cycles of reproduction, the strong, fast, and smart would drive out the weak, slow, and stupid. Differential reproduction would select those species members whose characteristics were best suited to the demands

imposed by the environment. The seemingly perfect fit that naturalists had observed between organism and environment was not the
product of intelligence or design—divine or otherwise. Rather, the
organisms that fit were the only ones that were left. The organisms
that fit had evolved, or been selected in evolution, precisely because
they did fit. Evolution was the survival of the fittest. And so the
modern blackbird and cowbird are the survivors. Some ancestor cowbirds may not have gotten rid of botflies. Some ancestor cowbirds may
not have laid eggs that looked like blackbird eggs. Some ancestor
blackbirds may not have tolerated botfly-destroying cowbirds. All these
ancestors are on the scrapheap of evolution.

Darwin largely worked out this revolutionary theory of evolution
by natural selection by 1837 or so, as documented by Michael Ruse
in *The Darwinian Revolution*. He knew that individual members of a
species differed from one another. He also knew that, from this range
of variation, a population of individuals possessing particular advantageous characteristics could be molded. He knew this from the successful practices of animal breeders. What he was missing was an
engine that would drive the selection of certain characteristics over
others in nature in the way that animal breeders did artificially on
the farm. Then, in 1838, Darwin read Reverend Malthus.

What he read was Malthus's *Essay on the Principle of Population*.
The Reverend Thomas Malthus was an economist who shared much
with Adam Smith but was less sanguine about the future of society
than Smith had been. The reason for his pessimism was simple: population growth was occurring and would continue to occur at a rate
that far outstripped the growth of available resources. Before long,
there would simply be too many people for society to feed, no matter
how productive it became. The inevitable result of population growth
would be competition for society's scarce resources. Some would do
better in this competition than others; the fit would survive while the
unfit starved. The only way out of this mess that Malthus saw was
the exercise of voluntary restraint; people would have to marry late
and have small families if the explosion of population was to be checked.

As a matter of empirical fact, Malthus was wrong, at least in the
short run. The rate of population growth slowed substantially in the
years following his pronouncement. But what was most important
about his argument was that it gave Darwin the engine of natural
selection. Life was a struggle for existence, a competition for scarce
resources. Only the fittest survived this competition. Creatures with
useful or adaptive traits won out—were selected—over creatures

without them. In this way, the fit between organism and environment was assured. In Darwin's own words:

> A struggle for existence inevitably follows from the high rate at which all organic beings tend to increase. Every being . . . must suffer destruction . . . otherwise, on the principle of geometrical increase, its numbers would quickly become so inordinately great that no country could support the product. Hence, as more individuals are produced than could possibly survive, there must in every case be a struggle for existence. . . . It is the doctrine of Malthus applied with manifold force to the whole animal and vegetable kingdoms.

Thus did Darwin derive a theory of the dynamics of the natural world from Malthus's theory of the dynamics of the social world. Malthus's social theory was false, but Darwin's natural theory lives on.

Darwin and Morality

One could not have been an evolutionist of any type in Darwin's time without being mindful of the moral implications of evolutionary theory. Evolutionary theory challenged received religious views about special creation (as it still does, at least to some fundamentalist "creationists") and was viewed by many as a serious threat to the moral fiber that held society together. Thus, evolutionary scientists were frequently at the center of moral debate, in a way that scientists rarely are in the modern world. And if evolutionary theory posed a moral threat in general, then Darwin's version, postulating competition among individuals for scarce resources, with the loser perishing, was especially gruesome. It conjured up images of a Hobbesian nightmare, a war of all against all. So Darwin had to have something to say about morality.

Among moral theories of Darwin's time, there were two leading candidates for a general theory of morality. One candidate might be called "moral sense" theory. Its argument was in essence that people had an immediate, intuitive, instinctive sense of right and wrong. They would act spontaneously and unreflectively in ways that on cool, rational reflection would be judged as moral. Thus, someone jumps into the river to save a drowning man not because he thinks about it and decides that it's the moral thing to do but out of immediate instinct. It just turns out that human instincts are, in fact, moral ones. The second candidate—much better known and more durable historically—was utilitarianism. The utilitarian offered a calculus for judging morality. Moral acts satisfied the "greatest happiness" principle, contributing to the greatest good for the greatest number. And moral people should actually enter into such calculation in determining the moral consequences of potential actions.

In *The Descent of Man*, published in 1871, Darwin presented his own, natural theory of morality. In it, he attempted, at a stroke, to unite the moral sense and utilitarian theories and to show that a theory of evolution by natural selection was not a challenge to morality. First, he acknowledged the presence of a moral sense and argued that it derived from the so-called social instincts—the instincts that guided sexual reproduction and parental care of young. As to the content of the moral sense, that depended on the concrete nature of the social instincts. Only some social instincts would lead to survival and reproduction by the organisms possessing them and would be selected in evolution. Which ones? Why, said Darwin, precisely the ones intelligent people would choose by rational, utilitarian calculation, because these are the adaptive ones. People are then utilitarian by their biological natures; their impulses about what is right correspond perfectly with their rational calculations of what is right. Darwin's account established not just the unity of moral sense and utilitarian moral theories but the unity of is and ought. The moral code people have is just the one they ought to have, with their success as a species as justification. Survival is the ultimate arbiter of moral conflicts.

Darwin's moral theory thus made a case that evolution could coexist with morality. But it also did another thing that may in the end be far more important: it brought morality squarely into the domain of nature. People behaved morally as a result of natural law. They behaved morally automatically. After-the-fact, utilitarian calculation might be used to justify a moral act, but it didn't cause it. Moral acts were caused by the same sorts of factors that cause the social behavior of insects, or the botfly-killing behavior of cowbirds. To understand morality, one must look to the science of instinct, not to the canons of reason. It is simply a part of human, biological nature to be moral, self-sacrificing, and sympathetic.

Darwin did not spend much time addressing himself to the implications for current social practices of the theory of natural selection. But other evolutionists of the time did. Notable among them was Herbert Spencer, who shared most of Darwin's views and even anticipated some of them and who turned them toward an examination of society. And there was plenty going on in society to examine. For example, in many parts of Europe, communities had passed poor laws—income redistribution plans in which workers were taxed to obtain revenues to support the poor. Some workers even received income supplements if their wage was below a given level. These plans were not, by and large, a glowing success. Some, for example, seemed to induce people to work less, decreasing their wage and thus increasing the public burden. Spencer looked at such social practices with the eye of a naturalist, not a moralist, and came to the conclusion that the government should not interfere, that redistribution was a mistake. It was a mistake not because it was immoral—not because people had the *right* to keep what they earned—but because it went against nature. It wouldn't work; no, it *couldn't* work. Buttressed by the principle of natural selection, Spencer argued for the invisible hand, laissez-faire, free-market system, by appealing to natural imperatives rather than moral ones. In so doing, he contributed to what came to be called social Darwinism—the application of Darwinian principles to social practices as a natural defense of entrepreneurial capitalism. This kind of relation between nature and society was just what economic science needed to sustain its claim that natural man was economic man. Biology and economics, the blackbird and the cowbird. The rest, as they say, is history.

Modern evolutionary biology has developed an even better fit with economics than Darwin could have imagined. It shares with economics the general view of organisms as maximizers of preference or utility. For economists, utility is very much an individual affair. Because different people want different things, a wide range of diverse economic activities will occur. But they will have in common the maximization of utility on the part of the individuals engaging in them. In contrast, for the biologist, the coin of utility is universal. Maximizing utility is maximizing reproductive fitness. As Peter Medawar has said, " 'Fitness' is in effect a system of pricing the endowments of organisms in the currency of offspring, i.e., in terms of net reproductive performance." The kind of activity that serves fitness maximization will vary from species to species, and even within a species, as environmental conditions vary. But all members of a species would be expected to make similar choices in similar circumstances.

Modern evolutionary biology also shares with economics the idea that the pursuit of interest is the pursuit of self-interest. Organisms are selfish. If acts of apparent self-sacrifice occur, rest assured that it is in the reproductive interest of the individual. If organisms act in ways that on the surface seem altruistic, rest assured that the motives are selfish. "Scratch an altruist, and watch a hypocrite bleed." Indeed for some, it isn't even the individual organism that is selfish, but individual genes that happen to reside in the organism's body.

Armed with these principles—of organisms (or genes) as selfish maximizers of reproductive fitness—the challenge to the modern biologist is to show how natural patterns of behavior in various species, some of which seem rather unselfish and unmaximizing, all exemplify the principles. And the new biologists have met this challenge with a cleverness and ingenuity that makes one take notice.

The Explanatory Framework

Modern evolutionary biology remains faithful to the key Darwinian insights that the evolutionary process depends upon the two ingredients of individual variation and selection. But these insights have undergone considerable refinement over the years. Before discussing particular examples of modern biological analysis, it will be helpful to sketch the general explanatory framework that guides evolutionary biology.

First of all, natural selection may operate on any characteristics of an organism that affect its ability to adapt to its environment. Most obviously, natural selection can work on features of an organism's structure, or morphology. Organisms can be selected for size, strength, lung capacity, and so on. An often-cited example of selection on the basis of structure is the case of a particular light gray moth that existed in abundance for thousands of generations in English forests. That it was light gray was not an accident; this coloration allowed it to blend in beautifully with the light gray tree trunks around it. Thus, its color provided camouflage. But then came the industrial revolution, and with it a perpetual cloud of black soot that darkened the countryside, including the trunks of trees. Now, all of a sudden, light gray moths stood out as clear targets for eager predators. Within a hundred years, light gray moths had virtually disappeared, having been replaced by black ones. More recently, with pollution-control measures cleaning up the sky somewhat, tree trunks are turning gray again—and so are moths.

This selection of moth coloration comes about not because some clever moth decides to have black offspring when it sees its fellows being spotted and eaten nor because black moths start accidentally arising when the trees turn black. Rather, the idea is that occasional black moths have always arisen (individual variation). When tree trunks were gray, these black moths were eaten before they could reproduce so that only the gray ones remained. When tree trunks started turning black, the selection contingency changed; now the occasional black moths were reproducing successfully while the gray ones were being eaten early. It didn't take long for black to replace gray as the dominant moth color.

Not only does selection work on aspects of structure, but it also works on behavior. It selects cowbirds that destroy botflies over those that don't. It selects good hunters over bad ones, fast runners over slow ones, parents who are effective at defending their young against predation over those who are not, and so on. One can demonstrate selection of behavior in the laboratory by doing selective breeding. First one measures some behavioral characteristic, say speed, in individual members of a species. If fast runners are then mated with fast, and slow runners with slow, over successive generations of selective mating, the two groups of animals will become increasingly divergent in their speed.

Among the patterns of behavior that can be selected is a particularly important class that will be the focus of much of this chapter. That class is social behavior. Patterns of interaction among members of a species are as susceptible to selection as anything else. Social behavior includes mating, caring for young, defending oneself against, and sometimes fighting with, other members of the species. Even elaborate patterns of social organization in large groups of animals are grist for the mill of selection. This is especially true among groups of insects, known as social insects, that live in large colonies in which individual members serve very specialized roles. Bees and ants are common examples of social insects. The study of how selection operates on social behavior is the province of a subdiscipline of evolutionary biology known as *sociobiology*. The concepts of sociobiology are of special significance when attention turns to the relevance of evolutionary biology to human nature, since social interaction is so obvious and fundamental a characteristic of all human life. If evolutionary biology can't provide a persuasive account of social behavior, its relevance to human affairs will be severely restricted.

So selection can operate on aspects of structure and on aspects of behavior. The next question that must be addressed is what is the unit of selection. The common wisdom is that the unit of selection is the individual organism. Well-adapted organisms survive and reproduce, passing their adaptive characteristics on to the next generation, while poorly adapted organisms fail. However, Darwin thought that selection could also operate on groups. A group that contained some self-sacrificing individuals might be better adapted than a group that did not. It would therefore produce more offspring, including some that were self-sacrificing. If there were competition among groups for limited resources, the better-adapted group would win. It is tempting to think in these terms about the success of some human societies relative to others.

In the human case, group selection may well be possible. Adults teach their young. Cultural practices are handed down from generation to generation. Human social groups with effective cultural practices may well be selected over groups with ineffective ones. But modern evolutionary biology tells us that group *biological* selection is virtually impossible. The reason is that the vehicle for transmission of traits, unknown in Darwin's time, has now been identified. It is the gene, a complex collection of protein that resides in every cell of every organism. If a characteristic is going to be passed on by one generation to the next, other than by learning, it has to be carried in the genes. Furthermore, the direction of transmission is one-way. Genes influence structure and behavior. Structure and behavior do not influence genes. That is, if an organism were to develop some trait in the course of its lifetime that proved extremely beneficial, it could not pass that trait on to its young in its genes. Acquired characteristics cannot be inherited. Giraffes did not grow long necks by stretching to reach leaves in tall trees. But even if they did, their offspring would not inherit their *acquired* long necks. This unidirectionality of influence—from genes to behavior—is sometimes known as the "central dogma" of evolutionary biology.

Given this central dogma, it is clear why group selection is so unlikely. Consider a concrete example. Certain species of birds utter "alarm calls" when they spot a predator. These calls warn other birds of the predator's presence, and they then take flight or shelter. These alarm calls clearly do the birds a lot of good. A group of birds containing some alarm callers would certainly be in better shape than a group without them. But what about the alarm callers themselves? It is not unreasonable to suppose that alarm calling is risky. At the same time that the alarm warns other birds, it alerts the predator to the location of a prey. Imagine the worst case; the alarm caller makes its call and immediately gets eaten. That's great for the other birds, but meanwhile, the eaten caller will not produce any children. And if alarm calling is carried in the genes, this means that the next generation will be deprived of some potential alarm callers. It is easy to see that over successive generations, alarm calling will disappear, even if the group is worse off as a result. Unless there is something in it (in terms of reproductive success) for the alarm caller itself, alarm calling will simply not be sustained by the group. Now were it to turn out that alarm calling was only done by birds that were past the age of reproduction, so that their self-sacrifice occurred *after* they had sired the alarm callers of the next generation, alarm calling would not be selected against in evolution. But even under these circumstances, there would still be nothing to select *for* alarm calling. Alarm calling could not be positively affecting the reproductive success of the caller if it didn't occur until the caller's reproductive activity was finished. No, so long as the central dogma is accepted, it is unclear how group selection could operate.

The argument against group selection has a very important consequence. It means that organisms are fundamentally, biologically *selfish*. They are out to maximize their own self-interest, meaning reproductive success. This is not because they are nasty; it is simply inherent in the logic of natural selection. Unselfish organisms who put themselves at risk for their fellows will not live long enough to produce unselfish offspring. Only the selfish will survive. It can't be any other way. But if selfishness is inherent in nature, it really does make behavior like alarm calling a mystery. How has it persisted?

The attempt to explain this seemingly unselfish behavior, and others like it, in the face of a logic that demands selfishness, has led some evolutionary biologists to the view that the unit of selection cannot be the individual organism. Rather, it must be something even smaller. It isn't organisms that are selfish but the genes that inhabit their bodies. Organisms are simply nature's way of making other genes.

This point of view is most forcefully articulated by Richard Dawkins, in his book *The Selfish Gene*. Think back to the time, says Dawkins, when there weren't any organisms, just chemicals of various kinds floating around in the primordial soup. First, different types of molecules started to form, by accident. Some of them, it turned out, were capable of reproducing themselves. That is, they took in constituents from the soup—carbon, nitrogen, hydrogen, oxygen—and turned them into copies of themselves. Now these various constituents were not in endless supply, so the different molecules were competing for them as they went about making copies. From this competition, the most efficient copy makers emerged. Furthermore, copy making was never perfect; sometimes the molecules made mistakes. And sometimes the mistakes turned out to be even better at making copies than their progenitors. One such mistake might have been the formation of a thin membrane that held the contents of the molecule together—a primitive cell. Another mistake might have involved dividing the cell into components, each with different characteristics. Another mistake might have involved clumping a bunch of these components together—a primitive organism. In each case, the mistakes proliferated because they were better at making copies of themselves than their predecessors were. And the best copy makers—the ones that have come down to the present—were the ones that happened to inhabit bodies.

If we think about the origins of life in this way, it is easier to see selfishness as inherent in the logic of the situation and not as a brutish characteristic of personality. Molecules weren't heartless cutthroats; they didn't *want* to proliferate. They simply did. They didn't *want* to inhabit bodies, they simply did. They didn't *want* these bodies to work in their interests; the bodies simply did. Those molecules whose bodies did not work in their interest have not survived. The surviving, "selfish," reproducing molecules are what we call genes.

Selfish gene theory can illuminate many characteristics of organisms that are otherwise puzzling from the point of view of natural selection. For example, it has been suggested that what is thought of as the natural process of aging and bodily decay—the wearing out of bodily organs after years of hard and good service—is actually the work of particular genes in the body. These genes are unusual in that they just lie dormant, having no discernible effect, for years and years. Then something triggers them into action, and the process of decay begins. How could such lethal genes persist? It is certainly not in the interests of the organism to possess them. The answer to this question has two parts. First, the genes can persist if their effects are deferred until after reproduction is over. Then, they would survive into the next generation before they began exerting their lethal effects. Second, the "interests" of the organism are irrelevant to the gene. The gene's "interests" are to reproduce itself, and the organism be damned. The interests of the gene and the interests of the organism it inhabits need not coincide, as is obvious in the case of lethal genes. Now lethal genes that have their effect early in the life of the organism, before reproduction has occurred, will not survive. They will be selected against. The pressure of selection will be to push the timing of lethal effects to later and later in the organism's life. Indeed, all organisms may be veritable time bombs, walking around with a whole collection of genes that are bad for them but that they will inflict on their children because the bad effects stay hidden until it's too late. The collective effect of all these bad genes may even explain why the capacity to reproduce diminishes as organisms age. The point here is that "bad genes" don't care, in any sense, that they're bad for the organism, as long as the harm they do does not interfere with their own reproductive success.

Selfish gene theory allows an explanation of the alarm calling of birds. The gene controlling alarm calls is interested only in itself, not in the body it inhabits. If it's bad for the bird, that's too bad, as long as it's good for itself. How can it be good for itself? Well, among the birds that the alarm caller saves will be some of its relatives. One of the important things about relatives (the only important thing as far as the gene is concerned) is that they are likely to have genes in common. So by making an alarm call, and sacrificing itself, a bird may be saving the alarm-calling genes of countless children, siblings, cousins, aunts, and parents. Reproducing itself is all the gene cares about, and if it can do this effectively by saving the genetic relatives of the organism it sacrifices, that's perfectly fine. This idea, known as the principle of *inclusive reproductive fitness,* is critically important to the explanation of many forms of social behavior. The point of inclusive fitness is that reproductive success is measured in terms of the success not only of oneself but of all one's genetic relatives.

Now that we are familiar with the view that it is not the organism, or its structure, or its behavior, that is the unit of selection, but the genes that determine that behavior, that structure, that organism, we can ask what exactly the biologist means when she says that the gene "determines" anything. What *is* the notion of genetic determination? People frequently talk casually of a gene for structure X or behavior Y, as if by poking around at the molecular level they could tag the particular chemical array that makes wings, or flying, or nesting, or fighting. This is inaccurate. Talk about genetic determination is really talk about the determination of *differences* between individuals. The emergence of wings or of flying depend upon a host of factors. They depend upon many genes, upon proper embryological development, upon proper nutrition, and upon the presence of crucial environmental factors. What genetic determination means is that with all these other factors held constant, differences in structure or behavior among members of a species can be traced to differences in genetic makeup. Thus, those who argue, for example, that there is a significant genetic component to human intelligence do not mean to suggest that if you take someone with "smart" genes and lock him in a room with little nourishment or social contact he will come out at the age of ten smart. No, smartness depends on a host of factors. The claim for genetic determination is that with all these factors held constant, the right genetic differences will lead to smartness differences.

People also tend to think that genetic determination implies complete inflexibility or unmodifiability. Aspects of behavior that are not genetic are easily modified by experience, by learning. Aspects of behavior that are genetic are locked in. This is not so. What genes may determine is tendencies, not actualities. Genes may make certain activities very likely and others very unlikely. But they needn't make them either inevitable or impossible. For humans to fly may in fact be genetically impossible, since they don't have even approximately the right anatomical structure. But it is not genetically impossible, even if rarely done, to run a mile in four minutes, or to run 26.2 miles without stopping. Thus those who argue that there are genetically determined differences in behavior between male and female people do not necessarily mean to suggest that it is impossible for a female to be very aggressive, or for a male to be nurturant. Rather, the argument is for a genetic *tendency* for males to be more aggressive than females and for females to be more nurturant than males. Though it may take work, these tendencies can be overridden by the right kinds of life experiences. The way in which genes control behavior is not like the way pulling the trigger of a gun controls the path of a bullet but like the way propelling an arm controls the path of a paper airplane. Tendency, not determination.

There is one last feature of evolutionary biological explanation that needs to be discussed. What, for the evolutionary biologist, does it mean to *explain* something? The biologist "explains" in two quite different ways. Consider, as an example, the question of caring for young by parents. Why does a wasp go hunting every day for food that it then places in the various holes in the ground in which it has placed its larvae? One approach to answering this question seeks immediate causal mechanisms. Is there some hormonal change in the wasp that triggers hunting? Is it the position of the sun in the sky? Is it the sight of appropriate insects? Is it the absence of food in the nests? Is it all of these things in combination? This approach does not distinguish the biologist seeking explanation from the physicist seeking explanation.

But the second approach does. For the biologist can also explain by appealing to the long-term, functional value of some behavior. The wasp hunts for food for its young because feeding the young promotes their survival, which in turn leads to successful reproduction. Behavior is to be explained in terms of its survival value, as well as in terms of the immediate mechanisms that control it. The rough rule-of-thumb is that if a behavior did not have survival value, it wouldn't be there— it wouldn't have been selected—and the puzzle facing the biologist is to figure out exactly what the survival value is. The idea is that behavior has been selected in evolution because it promotes the organism's inclusive reproductive fitness. Improvement in reproductive fitness is what characterizes evolutionary change; it is the raison d'être of evolution. And explanation that appeals to the improvement of reproductive fitness is what is really distinctive about evolutionary biology.

Sociobiology

While economic men may spend their resources on just about anything, from the point of view of evolution, the coin of the realm is reproduction. The function of being able to feed, shelter, or defend oneself is to have kids. All the activities that are directly involved in reproduction are social. Choosing a mate is social. Raising young is social. Protecting what one has from intruders is social. The branch of evolutionary biology that is specifically concerned with the study of these social activities is *sociobiology*. In the decade following the publication of E. O. Wilson's landmark book, *Sociobiology: the New Synthesis,* which essentially identified sociobiology as a coherent, new discipline, sociobiology has captured the public imagination. Having formulated principles that they think can account for animal social activities like mating, child care, and aggression, sociobiologists have not been shy to apply those principles to the same activities in humans. Examples include Wilson's *On Human Nature,* Dawkins's *The Selfish Gene,* and David Barasch's *The Whisperings Within.*

Sociobiology represents a provocative challenge to the ordinary way in which people understand their social activities. It attempts to show that the behavior of animals serves reproductive fitness by applying the economic notion of rational self-interest not to commodity pursuit but to social behavior—to sex, parenting, aggression, social interaction, and cooperation. More accurately, it applies the economics of commodity pursuit to social behavior by treating such activities as sex and parenting as commodities.

And the domain of social behavior provides the greatest challenge to evolutionary explanation, not just because it is central to reproduction, but because on the surface it seems to violate the key idea that organisms are selfish. We already encountered the apparent self-sacrifice of the alarm-calling bird. Like the soldier who makes a racket, calling attention to himself so that his comrades can escape, the alarm-calling bird is risking its life for the good of the group. Perhaps phenomena like this can be treated as isolated anomalies in an otherwise pristine picture of organisms in pursuit of self-interest: peculiar, hard to understand, but not so commonplace as to be bothersome. But the trouble is that such acts of apparent self-sacrifice are absolutely rife in nature. Every parent feeding and protecting its young is engaged in an act of self-sacrifice. Acts of parenting cannot be dismissed as "mere" anomalies. So a great deal hangs on the ability of sociobiology to paint a picture of social behavior that is consistent with the idea that organisms are out to maximize their own individual reproductive fitness.

In general terms, painting this picture requires thinking of selfish genes, not selfish organisms. Selfish genes are interested in more than just the organism they inhabit; they are also interested in that organism's relatives—children, parents, siblings, uncles—because copies of these selfish genes are likely to reside in those bodies as well. Maximization of fitness is the maximization of *inclusive reproductive fitness* (inclusive of all organisms likely to share one's genetic makeup.) Sociobiologists seek to explain the panoply of social activities that animals engage in by appeal to the pursuit of inclusive reproductive fitness.

Mate Selection and Sex

If the maximization of reproductive fitness is at the heart of evolutionary theory, then sex and the selection of mates is at the heart of evolutionary theory. But before talking about varied patterns of mating observed in different animals and the ways in which they can be understood to maximize reproductive fitness, something needs to be said about sex itself. For it turns out that sexual reproduction, which is the kind of reproduction with which people are so intimately familiar, is something of an evolutionary mystery. To see why, we need to understand some things about genetic material and how it is transmitted to offspring during reproduction.

Every type of organism has a fixed amount of genetic material, which is duplicated in every one of its cells—except for its sex cells, sperm for the male and egg for the female. These cells each have half the genetic material that the other cells do. This way, when they come together (the egg is fertilized), the offspring will contain the same amount of genetic material as its parents, by simple addition of the genetic material from the egg and the genetic material from the sperm. Exactly which half of the genetic material is contained by any particular egg or sperm is largely determined by chance (genetic material is organized into clumps called chromosomes, which are themselves arranged in pairs. The egg and sperm inherit one member of each chromosome pair). This characteristic of sexual reproduction means that the odds are fifty-fifty that any particular bit of genetic material will find its way into a particular member of the next generation. If some gene is on the chromosome that is left behind when the egg is formed, it's tough luck for that gene.

This is why sexual reproduction creates a problem, at least for selfish gene theory. Genes would make out much better if organisms reproduced *asexually*, that is, by simply making exact copies of themselves, with no partners (as some organisms do). Then *all* of the genetic material, not just half, would find its way into the offspring. In light of this seemingly enormous advantage to the genes of asexual reproduction, one must wonder how the genes could ever have let sex happen.

The puzzle of sexual reproduction does not have a clearly agreed upon solution. However, attempts to solve it are instructive because they reveal something fundamental about the way evolutionary biologists think. Maybe sexual reproduction is not so adaptive—not so fitness maximizing. Can't be, says the biologist. If it's there, and especially if it's so widespread, it *must* be adaptive. The question is, how is it adaptive?

Among the answers to this question that have been proposed, one suggestion goes like this. It isn't good enough simply to have offspring; the offspring have to live long enough to reproduce themselves. Living long enough will require food, water, and protection against predation. Now suppose that, when organisms mate, they also stay around to care for the young. And suppose that the whole of a mating pair is greater than the sum of its parts. That is, suppose that two adults can provide more than twice as much food, water, and protection than either could alone. Then, sexual reproduction would make good economic sense for the genes. A mating pair, let us say, could manage to support six offspring to reproductive age. This means that any particular gene is likely to be in three of them. If either mate alone could only support two offspring if it reproduced asexually, the genes make out better by participating in sexual reproduction.

There are problems with this clever argument, among them that sexual reproduction occurs even in animals that abandon their young completely right after birth. (Indeed, some animals abandon young *before* birth, when they are just fertilized eggs.) A second explanation of sexual reproduction is more persuasive. Sexual reproduction confers a significant selective advantage because it is the principal source of genetic diversity in a species. Since any offspring will inherit half the genes of each parent, the offspring's complete genetic makeup will be different from that of either of its parents. The new genetic combinations that occur with each act of reproduction provide the raw materials on which selection can operate. Some fortuitous combination of genes from its parents may make an animal bigger, stronger, faster, more skilled as a hunter, than any member of its species before. Furthermore, new gene combinations may be the primary device species have for coping with changes in the environment, for adapting to them over successive generations. Without sexual reproduction, genes would be stuck making carbon copies of the organisms they inhabit. Barring an occasional genetic mutation, there would be no mechanism to permit species change. Being a part of a system that is capable of change may be of such great survival value, even to individual genes, that it more than compensates for the fact that sexual reproduction confers only half of one's genes on any particular offspring.

The argument here should seem familiar. It is made again and again in defense of democratic guarantees of individual freedom of expression, movement, congregation, and so on. While in the short run such freedoms may be costly to a society (it makes concerted social action more difficult; criticism of aspects of society can create factionalism and undermine morale, and so on), the argument goes that, in the long run, it is beneficial. The exercise of these freedoms is the source of variability (in ideas, attitudes, actions), which allows a society to evolve and grow. So sexual reproduction is the evolutionary biological equivalent of freedom of speech.

Now that it is clear that a case can be made for the adaptive value of sexual reproduction, what about the particular patterns of mating that occur in different animal species? Much sociobiological work on this question has focused on differences between the sexes in mating patterns, especially in species most closely related to humans. How ought a man or woman out to maximize inclusive reproductive fitness to behave? The reproductive potential of men and women is quite different. A woman produces an egg once a month. Furthermore, once the egg is fertilized, her reproductive potential is finished for almost a year, and she is saddled with carrying the fertilized egg around, and protecting and nourishing it, for nine months. In short, a woman's reproductive potential is quite limited. In contrast, a man's reproductive potential is virtually infinite. He has a practically unlimited amount of sperm and can deliver them, on demand, day after day, week after week, year after year. A host of predictions about optimal reproductive behavior seem to follow from these differences between the sexes. They can be summarized with the phrase *parental investment.*

The female has an enormous investment in each egg. There will not be many opportunities for reproduction, so she wants to make the most of them. What this implies, first, is that she ought to be pretty selective about whom she mates with. She ought to seek a mate whose genetic contribution to the offspring is likely to make them reproductively fit. She may be looking for the biggest, strongest, smartest male she can find. Second, once she has mated, she ought to do everything she can to see to it that her fertilized egg makes it to birth and to adulthood. She ought to want to stay around and care for her young. And she ought to find to want a mate who appears willing to stay around and help out. Thus, women should be choosy about mate selection, willing to stay faithful to a mate if he helps out around the house, and nurturant, supportive, and protective of their young.

Now what about the male? His interests are very different. He has rather little investment in each sperm. He has little investment in each fertilized egg. In the time it takes for one fertilized egg to be born, the man can be fertilizing hundreds more. In the same time that it takes a woman to get just one genetic representative into the world, a man can produce a thousand. Furthermore, unlike the woman, a man can never be sure in any particular case that he is really the father. So a man might be expected to be rather *unselective* in choice of mate (after all, he can choose a different one tomorrow). He might also be rather unwilling to stay around and care for his offspring, when he could instead be out producing more. The male, in short, should be indiscriminate in choice of mate, promiscuous rather than faithful, and indifferent to the fate of his offspring. Does this little scenario about the different interests of the two sexes sound familiar? Well, the sociobiologists argue, it rests on rather firm biological foundations.

This does not mean that men and women will mate in a way that reflects an inflexible, slavish pursuit of reproductive self-interest. Remember, genetic determination is of *tendencies,* and they can be overridden. Indeed, the pattern just sketched is a caricature of actual relations between the sexes. But caricatures are not unrelated to the things they caricature; they may be exaggerations, but they are recognizable exaggerations. And so the sociobiologist would argue that if society wants to foster patterns of relations between the sexes that are different from the one just sketched, it must appreciate how deeply rooted these particular patterns are, and how hard it is for organisms, even people, to overcome their genetic "whisperings within."

In species other than humans, evidence that males and females have these different reproductive interests—and pursue them—is plentiful. It is commonplace for the females of the species to be the the selective ones in mate choosing. Males exhibit flashy coloration (at some cost, since it makes them conspicuous to predators at the same time that it makes them conspicuous to potential female mates), or display feats of strength and athletic prowess, to make the females take notice. Males will "court" females, putting up with a period of flirtatiousness and doing things for them like building nests or gathering food in order to win them over. Females are attracted to males who are willing to put in the time for nest provisioning, even before mating has occurred, perhaps because this time spent will increase the male's investment in the partnership, making it more likely that he will stay around when the offspring are born. Or failing that, at least he will have made a contribution to the care of the young before he runs off.

In some cases, the female's demand that the male contribute to her material welfare before she permits him access to her for reproduction is quite explicit. The female honeyguide, for example, is a bird that loves to feed on beeswax. She will offer her sexual favors only to males enterprising enough to control a bees' nest and willing to pay for sex with beeswax. In certain species of flies, the male's access to the female depends on his presenting her with a cache of food, wrapped up in a ball of silk. The male copulates while the female busily unwraps the package (a good thing too, for some of these females are very aggressive and are as likely to bite the male's head off as copulate with him; the silk-wrapped gift serves as a life-saving distraction).

There are examples in which female selectivity goes to extremes, leading to enormous competition among males for females. One such case involves the elephant seal. It is typical of elephant seal groups that a single male has access to all the females. He has a harem. The rest of the males are left out in the cold. The male who succeeds is typically the biggest, toughest male in the group. Succeeding at being number one is a genetic bonanaza; it gives the triumphant seal the chance to spread its genes throughout the next generation. But the costs of coming up short are equally dramatic. This leads male elephant seal pups to start taking risks in the service of being number one at a very young age. Male pups will attempt to steal milk by nursing at mothers other than their own. This is risky, because there is nothing in it for these foster mothers to allow foreign pups to nurse. After all, these pups will not be spreading the mother's genes. So the foreign mothers will attack milk thieves viciously if they spot them, often injuring them severely, and sometimes killing them. It's a big risk indeed, but one worth taking. For the male pup will have no reproductive future anyway unless he becomes the biggest and strongest adult. Interestingly, the female pup doesn't take this risk. There is no percentage in it for her; she will be able to reproduce whether she ends up big or small.

These patterns of sexual behavior, and many others that could be described, are devices that have evolved for resolving a fundamental conflict between the sexual interests of males and females, in terms of strategies that will maximize reproductive fitness. But in each case, one sex seems to be getting the better of the other. Sometimes males win the game, inducing females to serve their interests, and sometimes females win the game, but there always seems to be a winner. The tale sociobiologists tell about such games is that they are continuously evolving. The female wants a responsible male who controls important resources and will stay around and provide; the male wants to hit and run. The female won't mate with a male who looks irresponsible. It looks like the female has won this game. But then the male will evolve a strategy for deceiving the female, making her think that he will stay around, or that he has resources to provide. Thus, in one species of flies, the clever male gains access to the female by presenting her with a silk-wrapped package, but the package is empty; it has no food. It does the job, though. Why bother with substance when form will do. Now, the male is the winner. But only until the female evolves the means to read the deception, putting her once again on top. Of course, she will only be on top temporarily, until the male evolves a better deception, leading the female to evolve better deception detectors, and so on and on, in what Dawkins has called an "arms race" of thrust and parry, move and counter move.

When will this escalating gamesmanship stop? Perhaps it won't. The idea behind it is not that organism's *intentionally* develop better and more sophisticated devices for satisfying their own interests. Rather, a female who is easily fooled will be a reproductive disadvantage relative to females who can read deception. Over many generations, the more sophisticated females, able to spot males who will actually contribute to the care of the young, will come to dominate the population. This will put males who are unable to develop subtler deceptions at a reproductive disadvantage relative to males who can. There is no reason to suppose that this process—the evolutionary process—will ever stop. Unless, of course, one could evolve a deception that was simply unreadable. Then one would presumably have the upper hand permanently. It has been suggested by some sociobiologists that the best candidate for unreadable deception is *self-deception*. The male may not only act responsible and powerful, but he may do so because he thinks he really is responsible and powerful. He really expects to stay around and help care for the young. Then, they come, and much to his surprise (and his mate's) he bolts. How could she possibly have known when even he didn't. It looks as though the dictum "know thyself" is not an unmixed blessing, that self-deception can have adaptive value.

Alternatively, the arms race may eventually stop when both males and females adopt what are called *evolutionary stable strategies,* in which they deceive some of the time, are honest some of the time, win some of the time, and lose some of the time. Such strategies do not maximize the reproductive interests of either males or females. What they represent, however, is a compromise that cannot be improved upon by any other strategy that the bulk of the population might adopt. The important idea here is that of *frequency dependent selection.* How effective certain strategies are will depend on how common they are in the population. For example, if every male fly but one brought food to the female in their silk packages, then the one with the empty package would surely get away with his deception (unless he was individually recognizable). It would make sense for females to trust the males. However, if all the males were inclined to deceive, females would become distrustful. Deception, which was advantageous when unusual, would become disadvantageous when common.

To take a human example, consider something like paying income tax. Assume that taxes are used to provide services that people find useful. Should any individual (as an economic maximizer) pay his taxes? The answer is that as long as everyone else pays theirs, a particular individual should not pay his, assuming he can get away with it. After all, his tax is just a drop in the bucket. No services will be dropped without his few dollars. Thus, he can derive all the benefits of government services, without any of the attendant costs, if he just lets everyone else pay tax and tags along as a free rider. But the problem is that everyone else has the same idea. Everyone is out to maximize self-interest, not just him. So everyone wants to be a free rider. But if everyone is, government services will stop. What is good strategy for an individual, if he alone follows it, is not good strategy if everyone follows it. Thus, how good it is in fact depends upon its frequency in the population—frequency dependent selection.

The sexual analog of not paying taxes—say, not caring for young—may be best for individuals only if not everyone does them. Since everyone will try to do them, they will not be stable in evolution. The compromise strategies that evolve are the ones that lead to stable and positive results even if everyone follows them. Thus, almost everyone cheats on taxes a little, but almost everyone pays something. People pay more than they would like, but less than they are supposed to. Tax rates develop that are higher than they would have to be if everyone was honest, but just high enough to get things done by making allowances for cheating. These notions of evolutionary stable strategy and frequency dependent selection are very important. They will arise later, in connection with the sociobiological analysis of aggression.

The Sociobiology of Child Care

The female of a particular species of wasp spends virtually all of her time looking after her young. She first deposits her fertilized eggs in many separate little burrows in the ground. Then she supplies each burrow with food—caterpillars that she hunts and kills. After that, every morning, she inspects each and every burrow, determining whether its food supply needs replenishing. Having taken accounts, she spends much of the day hunting caterpillars, dragging them to the appropriate burrows, digging up the burrows, depositing the caterpillars, and resealing the burrows. She does this day after day until the young pass through the larval stage of development into adulthood.

If we think of the evolutionary dictum of survival of the fittest in terms of individual organisms, the mother wasp's behavior is simply incomprehensible. All that work, all that energy expenditure, and it does her no good at all. However, when survival of the fittest is thought of in terms of genes, rather than individuals, the mother's behavior makes sense. The point is inclusive reproductive fitness, and that can best be served by ensuring that as many young as possible make it to reproductive age. One's life is well worth sacrificing if it means the survival of several offspring. Thus, the very fact that parents care for their young at all challenges the idea that organisms are only and always out for themselves, at least in the sense in which we usually think about human selfishness. They are out for their genes.

Thinking about organisms as bodies out to maximize the interests of the selfish genes that inhabit them helps make sense of the fact that it is most common for the female of the species to make a greater commitment to child care than the male. The female member of the mating pair has a much greater investment in the viability of each offspring than does the male. She produces far fewer reproductive units than the male. She is stuck carrying them inside her body when they are fertilized. By the time they are actually born, she has already devoted a lot of time and energy to them. So she has a much bigger stake in their success than the male does. Thus, what is in the best interests of the mother is different from what is in the best interests of the father, with a variety of conflicts and compromises the result.

The same is true of parents and offspring. Their routes to the common goal of reproductive fitness are also different and can lead to a wide variety of parent–child conflicts. Each child has half of the mother's genes. So the mother should have an equal interest in the welfare of each of her offspring. She should not play favorites, and she should encourage her offspring to be cooperative and sharing with one another. But what about the kids? While each child shares half of its genes with its siblings, it shares *all* of its genes with itself. So each child is better served by being selfish and uncooperative than by sharing and cooperating. The picture of a distraught mother futilely pleading with her kids to stop fighting over every little thing and start being more cooperative should not be unfamiliar.

One of the situations in which this conflict is played out is when the mother feeds her young. Often, the mother's behavior toward her individual offspring will be governed by how hard they struggle for food. The nursing dog allows the scrappiest puppies access to her nipples, perhaps assuming that how much they scrap is related to how hungry they are. The mother bird deposits food in the beaks of those of her young who are cheeping the loudest. Now all would be well if the offspring quieted down as they got less hungry. Then, by always feeding the loudest complainers, the mother would end up satisfying everyone's needs. But it doesn't work this way. Kids cheat. They keep on complaining as loudly as they can, even as their need for food abates. It appears that they will take everything they can get. The noncomplainers end up becoming undernourished, which weakens them, which makes them complain even less. They become runts of the litter, often dying. This is especially true when resources are relatively scarce and the mother barely has enough food to go around.

It has even been suggested that this child-complaining for food is a form of blackmail. As the baby bird cheeps, louder and louder, it is essentially saying, "Here we are predators; come and get us." The mother has no recourse but to shut it up by stuffing it with food. Nest predation would be a genetic catastrophe for the mother. Thus, the spoiled baby gets whatever it wants. This blackmail reaches a high art in the case of the cuckoo. The mother cuckoo bird deposits her eggs in the nests of other birds. When the baby cuckoo hatches, it cheeps louder and longer than any of the mother's actual offspring. The mother is helpless. Ignoring the cuckoo puts her own brood at risk. And she can't really call the young cuckoo's bluff, because if a predator comes and devours the nest, it will not be the cuckoo's genetic relatives who get eaten. The mother has much more to lose in this poker game than the baby cuckoo does.

It may not seem terribly adaptive for the mother to be so insensitive to the actual needs of her young and to respond so slavishly to their protests. But appearances can be deceiving. Imagine a mother that is able to give to each according to its needs, in a situation of scarcity. There isn't quite enough to go around, but the mother makes sure that each child gets an equal share. The result of this behavior may be that none of her offspring survives to adulthood. If in contrast, she provides unequally, according to complaint, she may be sure to gain some surviving adults, even if it means sacrificing some of the litter early. Indeed, this device may serve to keep the local population of a species at a size that its environment can support. When food is abundant, everyone survives (even the most persistent complainers eventually shut up and give others a chance). When food is less plentiful, the complainers don't shut up, and one runt offspring is sacrificed while the others survive. When food is still less plentiful, two runts may be sacrificed; then three, and so on. A pattern of child care like this will work to match the size of the adult population to the available environmental food supply. So even if the mother did evolve the means to tell whether her offspring were cheating, it would not obviously be in her reproductive interests to use it.

The Sociobiology of Altruism

The discussion of the parent's willingness to make sacrifices for its child leads to the question of altruism more generally. Remember that our discussion of sociobiology began with the example of the alarm-calling bird, whose behavior appears to be anything but selfish. And remember Darwin's discussion of the moral sense that supposedly arises out of the social instincts. Darwin might have been wrong when he explained moral acts in terms of group selection, but he was not wrong in identifying them as acts in which either the individual suffers for the good of the group or the individual forgoes purely selfish, competitive, seemingly optimal behavior and cooperates with other members of the group. Acts of altruism or cooperation occur all the time, and they are not restricted to parents and their young. Many animals hunt for food in groups and share the catch. Worker bees sting intruding honey stealers and, in so doing, tear their insides up. The question that must be asked about altruism in general is, What's in it for the selfish genes?

This question has two different answers. One of them, already encountered in the case of parent–offspring relations, involves the concept of *inclusive* fitness, or what is sometimes called *kin selection*. The other, not yet encountered, involves the concept of *reciprocal altruism*.

Parents, siblings, grandparents, aunts, uncles, cousins, great-grandparents are all related to us genetically. As the relation grows more distant, the amount of genetic relatedness decreases (for example, one shares, on average, half his genes with each parent, a quarter with each grandparent, an eighth with each first cousin, and so on). Based on the theory of kin selection, acting altruistically will make sense if the beneficiaries of altruism are genetic relatives, and how much sense it will make depends on how closely related they are, and how many of them are benefited by the altruism. Suppose a man had to decide whether or not to jump on and smother a hand grenade, killing himself for sure, but saving everyone else nearby. Should he or shouldn't he? If he runs and hides instead, he will save one genetic unit—himself. Suppose that by smothering it, he saves three sisters. This represents a net gain in his genetic material, since each of his sisters possesses half his genetic material (thus, a total savings of one and one-half genetic units). But it wouldn't pay to sacrifice himself for three cousins. For self-sacrifice for cousins to pay, he would have to save at least nine of them.

The otherwise peculiar phenomenon of menopause can be understood as an act of altruism in this way. Menopause doesn't seem to make much genetic sense. It would make more sense for females to be reproductively capable for their whole lives. Well, suppose a point comes when the female is too feeble both to carry young and provide for them into adulthood. Suppose such a reproducing female would produce only one or two viable offspring in each litter. If instead of reproducing, she contributes to the care of her grandchildren and is responsible for the success of seven or eight in each litter, she comes out ahead genetically. Eight one-quarter relatives is a lot better than two one-half relatives.

This kind of calculation, of inclusive reproductive fitness, broadly construed, is how sociobiologists explain many instances of altruism. But calculation is used advisedly here. No one is suggesting that animals actually go around computing the degree of genetic relatedness of the organisms they will be saving. Altruistic acts can be quite automatic and uncalculated. However, the sociobiologist predicts that they will only persist in a population to the extent that other mechanisms work to ensure that the creatures saved will be genetic relatives.

Suppose, for example, that creatures tend to congregate with their relatives, and that their altruistic acts will benefit whatever animals are closest to them. In this case, altruism will be genetically advantageous. If the young of a species tend to scatter far afield of their relatives, altruism would not be genetically advantageous, and it would not be expected to persist in the population. There is much evidence consistent with this picture. Altruism tends to occur only in species where relatives stay together. Still more impressive, in some species, the females stay close to home while the males venture out on their own. In these species, the females, but not the males, are likely to engage in altruistic activity.

This raises other questions. How does a creature "know" that the animals near it are its relatives? How does it know that it is saving its own genes when it does something altruistic? First, the individual doesn't have to know. It can behave altruistically in blithe ignorance of the relatedness of its neighbors and still make out well genetically, as long as its neighbors *are* relatives. Second, there is ample evidence, in many species, of the ability to recognize kin, sometimes on the basis of common experience as littermates and sometimes apparently innately; sometimes on the basis of visual similarity, and sometimes on the basis of smell. In some cases, an animal that will behave competitively in a given situation to a nonrelative behaves cooperatively in the same situation to a relative. Thus, the evidence for kin selection as a mechanism underlying altruism is impressive.

But it is not the only mechanism. Sometimes altruism occurs among nonrelatives. Here, the sociobiologist appeals to reciprocal altruism, or, "You scratch my back, and I'll scratch yours." The idea behind reciprocal altruism is that it serves one to help an unrelated species member if one will be helped by that member in return. Indeed, the concept of reciprocal altruism extends between species. The example of the botfly-killing cowbird that began this chapter is an instance of cross-species reciprocal altruism. The blackbird feeds the cowbird (altruism 1) and the cowbird protects the blackbird from parasites (altruism 2). Such cross-species examples of reciprocal altruism are often referred to as *symbiosis*. And it should be understood that symbiosis is not an act of kindness; it is an act that is purely in the genetic self-interest of each of the symbiotic partners. One may be disinclined to view "altruistic" acts that are self-serving down deep as "genuine" examples of altruism, reserving the term for acts that entail real costs to the altruist. The sociobiologist's reply is that such acts probably don't occur at all. If organisms with a tendency to engage in them should happen to be produced, their genuine self-sacrifice will ensure that they don't pass their charity on to the next generation. Remember, "scratch an altruist, and watch a hypocrite bleed."

If acts of altruism are really selfishly motivated, are they really the very best things for a selfish organism to do? Suppose one lives in a small town in which a spirit of mutuality reigns. People look out not just for themselves but for their neighbors. Everyone is always willing to lend a hand. Each individual will surely benefit from the care and concern of others, but at a price. People will have to lend their hands to the general welfare also. Now what if they didn't? What if they derived the benefits of this community's altruism but refused to pay the price? What if they were free riders? Wouldn't they be even better off it they let everyone else do the work?

The answer is maybe. If the support of others was unconditional, that is, if they continued to assist people even without the prospect of reciprocation, an individual would be better off being a free rider. Or even if the community was not willing to offer unconditional assistance, if one could free-ride without being detected, he would also be better off. But now think: if one person were clever enough to get the idea to be a free rider, why can't someone else? And if one person were clever enough to get away with it, why can't others? Suddenly, there are two free riders. And soon, there will be three, four, twenty, a hundred, until the whole character of the community has changed. Now, with each person doing the individually "optimal" thing, everyone is worse off then before.

We encountered reasoning like this before when we talked about frequency dependent selection and evolutionary stable strategies. The idea is that pure, unconditional altruism is not evolutionarily stable. If a free rider should appear in the population, it will have an enormous advantage over the "suckers." Over the course of generations, this advantage will be played out in the form of greater reproductive success, until there are no unconditional altruists left. But a population of free riders is also not stable. If a few altruisists arise, and can recognize each other, and only help those who help them in return, they will have an advantage over the free riders. Over many generations, free-riding will disappear. However, if this mutual and conditional altruism is replaced by unconditional altruism, it will again be displaced by free-riding. Thus, while neither unconditional altruism nor unconditional free-riding will be stable in evolution, a pattern of conditional altruism can be. It can represent a strategy that cannot be improved upon by any other that is adopted by the bulk of the population.

Aggression

There is one last type of social behavior that has commanded the sociobiologist's attention. We have already seen that there are conflicts between males and females, between parents and young, between different males who are competing for a given female's attentions, between altruists and free riders. Are all of these conflicts resolved peacefully, in the way civilized human beings resolve them? Do animals sit down and work it out until the conflict is resolved? The answer is no. Sometimes these conflicts lead to fights, to aggression.

To understand when aggression occurs, and when it doesn't, and what patterns of aggressive behavior are in the best interests of the genes inhabiting a particular body, we must depend again on the ideas of frequency dependent selection and evolutionary stable strategies. Imagine two animals contemplating a fight (for, say, control over a food resource), and consider the potential costs and benefits of different actions. Each animal has two options: it can fight all out, to the death, winner take all, or it can genteelly threaten. Let us call the first option "hawkish" and the second "dovish." Now suppose animal 1 is hawkish. If animal 2 is a dove, animal 1 will win the encounter quickly. If animal 2 is a hawk, there will be a fierce battle. The loser will lose big, but the winner will also lose—time, energy, and personal injury. Presumably the benefits of the gained resource will be worth more than the costs in fighting (else why fight over the resource at all), but the costs will be substantial. So a hawk will quickly and easily vanquish a dove and will slowly and painfully vanquish another hawk. What about when dove faces off again dove? Here, both will dance around pretending for a while. Eventually one will win, but at little cost in confrontation to either.

Should animals be hawks or doves under these conditions? To help answer this question, points can be assigned to the various possible outcomes of encounters. A victory is worth 100 points, a loss is worth no points, serious injury is worth -200 points (has a cost of 200 points) and time spent threatening is worth -20 points. Suppose that when like meets like, it has a fifty-fifty chance of winning the encounter, but that hawks always beat doves. If a population began with all doves, the outcome of the average encounter would be computed as follows: when a dove wins, it gets 100 points minus 20 points for the time spent, or 80 points. When it loses, it gets -20 points for time spent. Since it wins half the time, its average score is the average of winning and losing, or $(80 - 20)/2$, or 30 points.

Now let a hawk appear. Since it will always be fighting doves, it will always win. Since doves won't put up a fight, it will never experience any cost. Thus, the hawk will gain 100 points in every encounter. What an enormous advantage. Hawks should spread like wildfire through the population, competely dominating doves. Thus, an all-dove population is not stable.

Suppose the hawks really are so successful that they take over completely. How will encounters go now? When a hawk wins, it will gain 100 points; when it loses, it will lose 200 points (remember, hawks fight tenaciously). On average, winning half the time, the hawk will end up with -50 points. Now let a single dove appear. Its score will always be 0, since it loses to all hawks, but it can just bide its time on the sidelines while the hawks kill each other off (0 is better than -50). Slowly but surely, dove genes will spread through the population. Thus, an all-hawk population is not stable.

What is stable is a mixture of doves and hawks, or a uniform set of animals each of which is sometimes hawkish and sometimes dovish. Even better is a strategy that is not simply randomly hawkish and dovish but depends upon the behavior of the adversary. Such a strategy, which might be called a "conditional strategy," has an animal start out being dovish, just threatening. If its adversary is also dovish, it stays dovish. But if its adversary attacks, it turns hawkish.

We have now reached the end of the sociobiological story. It is the updating of Darwin, the second coming of natural Malthusianism. It sees organisms as entirely selfish and economic, and selfish and economic by natural law. Indeed, it sees organisms as even more economic than economists do. For economists, with few exceptions, acknowledge that some human social relations lie outside the economy. The family is one of them. For the sociobiologist, *all* social relations are economic. Sociobiology extends economic analysis to domains like sexual and family relations. Sociobiology pins the kinds of relations observed between sexes and generations, the roles and strategies that are a familiar part of male–female and parent–child interaction in modern society, on biological rather than cultural determinants. It does not suggest that such roles and relations are inevitable. Societies could choose to restrain or subvert the biological impulses of their members. But they will have trouble if they ignore the biological "whisperings within." Nor does sociobiology suggest that such roles and relations are good. Dawkins says: "I am not advocating a morality based on evolution. I am saying how things evolved." And Barasch says: "Evolution is—or better yet, evolution does. It says nothing whatsoever about what ought to be." Thus, sociobiology tries to defend its status as morally neutral science, simply uncovering facts with which people may do as they please.

But if sociobiologists are right, what sense does it make to preserve this moral neutrality? What is the force behind "people may do as they please?" How can this discretion and self-determination coexist with the "human economic nature as natural law" flavor of sociobiology? One feels a little uncomfortable with the nuclear engineer who says, "I just provide the technology; it's up to everyone to decide how to use it. I bear no special responsibility for the moral consequences of nuclear energy." But at least there is nothing in the nuclear engineer's science to undermine the notion that people do bear moral responsibility. The sociobiologist, in contrast, is plying his trade right in the moral domain. If a given act is largely biologically determined, it makes little difference how people talk about it in terms of moral responsibility. If it makes a difference to talk about responsibility and discretion, then some of the force of the biological argument is undermined. Either people blame the river for overflowing its banks and flooding their houses or they look for natural laws that explain its behavior. One can't have it both ways.

Darwin seemed to know this, though his sociobiological descendants have apparently forgotten. Recall that Darwin's moral theory was one that made a virtue of necessity. People possess a moral sense—moral instincts—and they possess the particular moral sense they have because it has adaptive value. The possessors of maladaptive moral senses, however moral they may have been, have left no descendants. Darwin saw that if people are to take biological determination seriously, they must end up with a morality based on evolution. And so it is with sociobiology, the many disclaimers notwithstanding. This is the way it is; the question we must address is whether this is the way it should be.

use their past experience as a guide to successful (intelligent, fitness-maximizing, preference-maximizing) future action. And it is provided by the branch of psychology known as *behavior theory,* a discipline given its shape and direction by B. F. Skinner.

AN EXAMPLE

A pigeon that has been deprived of food is confined in a small, rectangular chamber. A dim ceiling light illuminates the chamber, which is bare except for two openings in one of the walls. Behind one of the openings, at roughly the pigeon's eye level, is a brightly lit plastic disk, mounted flush with the wall. Several inches beneath the disk is a rectangular hole behind which, out of the pigeon's sight and reach, is a food reservoir that contains feed grain. The pigeon moves about, seemingly at random, exploring the chamber. Occasionally, it pauses to groom itself. Suddenly, there is a loud noise. The food reservoir has been moved to a position near the rectangular hole, close enough to the wall that the pigeon can get at the grain. Initially, the loud noise seems to frighten the pigeon; it stays immobile, far away from the food. But eventually its fear diminishes, it approaches the food source, and it begins to eat. After a few seconds, the food is again made inaccessible. A minute or so later, the food is reintroduced. This time, the pigeon is much quicker to begin eating. Again, after a few seconds of eating, the food is removed, only to be made available again a minute or so later. This business continues, with food presented for a few seconds every minute or so, until the pigeon begins approaching the feeder as soon as it hears the sound, perhaps as a pet dog comes scampering into the kitchen when it hears the rustle of its bag of dog food being opened.

The pigeon has now been trained to eat from the feeder. At this point, the lit disk becomes important. Instead of food being delivered every minute or so, no matter what the pigeon does, the pigeon will be required to earn its keep, by pecking at the disk. The pigeon moves about the chamber, periodically sticking its beak through the feeder hole (to no avail). Then, by chance, it orients its beak in the direction of the disk. The feeder moves into position for a while. Now, when the food disappears, the pigeon moves straight to the disk, and seems to be staring at it. Again food is presented. Back to the disk, moving from side to side, bobbing its head up and down, but always oriented to the disk. Again food comes. When the pigeon goes back to looking at the disk the next time, its agitated movements are accompanied by a peck at the disk, which produces a resounding click. Success! More food! Back to the disk for another peck (along with its head bobbing and side to side movement) and more food. More pecks, and more food. As this process continues, much of the pigeon's extraneous movement slowly drops out. It seems to be learning that bobbing up and down and moving from side to side are beside the point; all it has to do is peck. The pigeon becomes an efficient pecking machine, moving up and down between the disk and the food like a yo-yo. The pigeon has been trained, or conditioned, to peck the disk for food.

Now, things change again. The pigeon pecks the disk, but food doesn't come. It pecks again, and again, and again, but still no food. It continues to peck furiously, but nothing happens. Gradually, its pecking diminishes. It goes back to wandering about the chamber, and grooming, and makes only an occasional peck at the disk. Because pecking no longer brings food, it has been eliminated or extinguished.

With the pigeon now pecking very infrequently, the relation between pecking and food is reestablished. The pigeon pecks (one is tempted to say half-heartedly), and food comes. Instantly, the pigeon returns to its yo-yo mode of activity.

Time for another change. The pigeon pecks, and food doesn't come. It pecks again, and again, with no luck. Just as it is ready to give up, it makes one last peck—its tenth—and food comes. Back to the disk it goes, and after another ten pecks, food comes again. So pecks are effective in producing food. It's just that now, instead of food for every peck, it is food for every ten. As the pigeon gets used to this ten pecks for every feeding ratio, it is increased—to twenty, then forty, then seventy-five, and eventually one hundred. The pigeon adjusts to each change, until eventually, the pigeon is pecking thousands of times in the space of an hour to obtain its daily ratio of food. The pigeon is working on a piece rate schedule, being paid with a few seconds of food for every one hundred pecks.

The Theoretical Framework

This little example, which has been repeated thousands of times in the laboratories of behavior theorists over the last forty years, embodies virtually all of the essential components of a behavior theorist's account of human nature. When the pigeon is first placed in the conditioning chamber, its behavior is highly variable. It does a wide variety of things, some of which are genetically determined and some of which are the result of the pigeon's past experience . Some of these things happen to be followed by a good or valued outcome—food. As a result, these things are *selected;* they occur with increasing frequency, occupying more and more of the pigeon's time. This fact, that behavior that is followed by valued outcomes tends to increase in frequency, is known as the *principle of reinforcement.* The idea is that valued events, like food, strengthen or reinforce the behaviors that precede them. In similar fashion, behavior that is followed by bad outcomes, for example, painful electric shocks, will decrease in frequency. This is known as the *principle of punishment.* Behavior will also decrease in frequency if it *stops* being followed by good outcomes. This is known as the *principle of extinction.* Because food followed pecking in our example, pecking became more frequent. When food stopped following pecking, pecking became less frequent. When once again, food followed pecking, pecking grew again in frequency, even when most of the pigeon's pecks were not followed by food.

It should be transparent that the principle of reinforcement is just the theory of natural selection brought down from the evolutionary history of a species to the life history of an individual. Natural selection requires (1) a source of variation, (2) a mechanism of selection, and (3) pressure to direct the selection process. Variability is genetic, the selection mechanism is differential reproduction, and the directing pressure comes from competition for scarce resources. Likewise with the principle of reinforcement. Organisms are characterized by endogenous behavioral variability. The pigeon starts out with a wide range of varied possible activities that may be thought of as competing with one another for the pigeon's time. The selection mechanism is the principle of reinforcement. The principle of reinforcement *selects* from the behavioral pool the particular activities that work. What it means for an activity to work is that it produces or is followed by some state of affairs that the pigeon "likes" or "wants" or "needs." And these likes or needs are what give the process of behavioral selection direction. As these "successful" activities are selected, they crowd out other, less successful, or unsuccessful activities. Only successful activities survive in the competition for the pigeon's time. Thus does the principle of natural selection—what Skinner calls "selection by consequences"—operate on both an individual and species level. Indeed, according to Skinner, it operates at a third level as well. The development of cultural practices can also be understood in terms of the selective effects of their consequences for the culture.

There is nothing terribly startling about the principle of reinforcement. People have surely known about it for millennia. A woman goes to a restaurant for the first time. If the meal is good, she returns.

If not, she doesn't. A man goes to a party and meets strangers. If he has a good time with them, he seeks them out in the future. If not, he doesn't. An employee makes a bold recommendation to a superior at the office. If she admires the initiative, the employee continues to speak up. If not, the employee goes meekly about his business. The examples can go on and on. The central idea is simple. Bits of behavior occur. Only the ones that produce beneficial consequences continue to occur. The others fade away. In this way people come to possess an effective repertoire of actions as adults. Flexible behavioral raw materials get molded, shaped, selected by the principle of reinforcement, so that of all the different things people might possibly do, they actually do only what works—or what has worked in the past.

The principle of reinforcement ensures that people continue to engage in only those activities that work—activities that get them what they want. But suppose that the things people want are bad for them. Suppose they want nicotine, alcohol, cholesterol, heroin. There is nothing about the principle of reinforcement to prevent people from following a direct path to self-destruction. If the things people want are bad for them, the principle of reinforcement will work to lubricate the slide into oblivion, by making people very efficient at getting those harmful things. Here is where the principle of reinforcement and the principle of natural selection work nicely together. One could imagine that people will differ in the things they want. Some will want the bad things while others will want the good ones. The principle of reinforcement will make everyone good at getting what they want, but the result will be that some people will destroy themselves before leaving offspring. Over many generations, natural selection will select people whose wants are not self-destructive.

People *know* that drugs, alcohol, fats, and tobacco are bad for them. Yet many persist in seeking these things. How can an explanatory framework, combining natural selection and reinforcement, emphasize the adaptiveness of behavior, when people continue to behave in ways that are patently maladaptive? Actually, if anything, the continued pursuit of harmful things makes the selection / reinforcement framework even more plausible than it would be if people were model citizens. To see why, it will be helpful to contrast the way people explain their actions in everyday language with the way the selection / reinforcement framework might explain those actions.

When we set out to explain someone's actions, we make certain assumptions about how people operate, among them that people are intelligent and rational. This involves assuming that people have *reasons* for what they do. They establish some goal, or desire. They formulate a plan for achieving that goal, and they execute the plan. Formulating and executing effective plans involves knowing something about how the world works and being able to anticipate the consequences of one's actions. Thus people act with foresight, using their knowledge to design plans of action that will realize specific aims.

Now plenty of what people do doesn't fit into this framework. People have no plan when they blink their eyes. They have no plan when they fall down after having been hit by a car. But behaviors such as these which are not planful and intelligent don't count as actions. It wouldn't even occur to anyone that such behaviors require an explanation. And of course, plans don't always work. People don't always achieve their desired goals. They make mistakes. But even mistakes can be understood by appealing to intentions. Thus, when a friend takes a long and roundabout route to work, we say, "She must have thought they were still doing work on the highway." Or when a colleague fails to appear at a meeting, we say, "He must have thought it was scheduled for tomorrow." It is when such purposive accounts of people's actions—even accounts that assume that people are making mistakes—cannot be constructed that people are regarded as prime candidates for psychiatric care.

The principle of reinforcement could certainly be understood from within the framework of everyday, purposive explanation. People want things (reinforcers). They figure out what has to be done to get them. And then they do those things. But from within a framework like this, one is really at a loss to explain the pursuit of alcohol, drugs, and other harmful objects. If a person knows they are harmful, and is rational, why does he pursue them? Perhaps, one might say, they are only harmful in the long run; in the short run, they are pleasurable. But surely if a person has foresight, the long run will not be lost on him. Rational, purposive people should not be pursuing harmful goals.

The major message of behavior theory is that while one *could* translate the principle of reinforcement into everyday, purposive language, it would be a mistake to do so. The principle of reinforcement produces intelligent behavior (that is, behavior that results in desired states of affairs), but it does not depend on human intelligence (foresight, planning, etc.) to do so. Instead, the principle of reinforcement should be thought of as an unintelligent mechanism whose product is intelligent action. The pigeon doesn't peck *because* it wants food and it knows that pecking will bring it food. The pigeon simply pecks, along with doing lots of other things. Pecking results in food, while the pigeon's other activities do not. Thus, pecking is selected and increased in frequency, as other, ineffective activities drop out (are extinguished). The end result is just what an intelligent organism would have done, but it is accomplished without the aid of intelligence as it is ordinarily understood.

The role of the principle of reinforcement in supplying an unintelligent mechanism to account for intelligent action is exactly paralleled by the principle of natural selection. Among naturalists of the eighteenth and early nineteenth centuries, one of the most compelling arguments for the existence of God was the presence of so many diverse forms of life, each of which seemed ideally suited to its environmental niche. Each creature seemed *designed* for the kinds of problems it would face in its environment. The fit between animal and environment was perfect, no matter what kind of animal one looked at. It seemed to these naturalists most unlikely that garden slugs, worms, cockroaches, and the like were intelligent enough to design themselves and their behavior to fit in so well in their environments, or to seek out just the environments to which they were well suited. Yet the fit was too perfect, and repeated itself too often, to be attributed to luck. Thus, it must have been that God had been the source of intelligence, designing creatures so that they would have the equipment that they needed to get by.

What natural selection did was take intelligence and design out of nature. Sure, every animal type was perfectly suited to its environment. But that was because nature's millions of mistakes were no longer around to be seen. Nature consisted of random variation in the characteristics of its creatures. Only those creatures possessing variations that enabled them to survive and reproduce remained to be observed. Nature's mistakes (ineffective variations) were buried in the distant past. So the intelligence or design that seems to be a part of nature is actually the result of an selection process that works to keep only those unintelligent (random) variations that are most effective. And so the intelligence or design that seems to be a part of human action is actually a result of a selection process (the principle of reinforcement) that works to keep only those unintelligent (random) variations in behavior that are most effective. Thus, for behavior theory, the principle of reinforcement represents not just a new fact (and not an especially startling one) to be added to our collection of facts about human nature. Instead, it represents a radically new and different way to understand human nature in general.

One of the virtues of the behavior theorist's framework is that it captures not only what seems to be intelligence in people but also flexibility. Patterns of behavior, once selected by the principle of reinforcement, are not written in stone. If the contingencies change, so that old patterns of behavior stop paying off while new ones start paying off, behavioral repertoires adapt to the new contingencies. Behavioral repertoires continue to develop, to evolve, meeting the changing demands imposed by the environment. Just as the direction of evolution can change if the environment changes, so can the direction of behavior. And just as in evolutionary theory, the responsibility for the paths that evolutionary changes take is placed not in the hands of God, nor in the foresight of evolving species, but in the selecting effects of the environment, so it is with behavior theory. Environmental contingencies and consequences—not foresight, intelligence, planning, or reason—are responsible for the paths that behavioral evolution takes.

So the selection/reinforcement framework really is very different from the everyday explanatory framework people are used to. Furthermore, the difference actually makes it easy for behavior theory to explain why people seem to want things that are bad for them. Distinguish first between the immediate consequences of something and its delayed, or cumulative consequences. The long-term consequences of, say, smoking cigarettes are bad—lung cancer, vascular disease, and who knows what else. A particular person starts to smoke cigarettes. Why does she do it? Perhaps smoking is pleasurable, perhaps it is relaxing, perhaps it is a sign of adulthood and independence, or a sign of daring, perhaps it serves as a substitute for snacking and is thus an effective weight control agent. For any or all of these reasons, the immediate effect of smoking is positive; smoking is reinforcing. One of the central principles of behavior theory is that the effectiveness of some outcome in controlling behavior is largely determined by the temporal proximity between outcome and behavior. The closer together in time response and outcome occur, the more effective the outcome will be. Thus, if a pigeon received food for pecking two minutes after it pecked, instead of right away, pecking would surely never come to dominate its activities. If the immediate effect of smoking is positive, this immediacy may dwarf the long-term negative effect, even if the negative effect is *very* negative. It is as though one added to the pigeon experiment described above a contingency whereby every five thousand pecks would produce an extremely painful electric shock. Do not suppose that the pigeon would stop pecking for immediate food because of this long-term negative contingency.

So the behavior theorist's answer to the question, why do people seem to want bad things? makes no appeal to rationality or irrationality, knowledge or ignorance. It appeals instead to the principle that immediate positive consequences can dominate delayed or long-term negative consequences in controlling behavior. Rationality is simply beside the point.

Central Concepts of Behavior Theory

The principle of reinforcement expresses a relation between events belonging to two categories: the organism's responses comprise one category; and rewarding and punishing environmental stimuli comprise the other. To these must be added a third category: environmental stimuli that are neither rewarding nor punishing. What can be said about the members of each of these categories and about the relations between them?

Since instances of responses and reinforcers were included in the example of the pigeon pecking a disk for food, let us begin with the third category—environmental stimuli that are not rewarding or punishing. Suppose the pigeon experiment is modified. Sometimes the disk is lit with red light and sometimes it is lit with green light. If the pigeon pecks the disk when it is green, food comes. But if the pigeon pecks the disk when it is red, no food comes. What does the pigeon do under such circumstances? It rather quickly learns to peck the disk when it is green but not to bother when it is red. Red means stop; green means go. The pigeon has learned a *discrimination,* not between red and green, for the pigeon already could tell the difference between red and green, but between the consequences of pecking the disk when it is red and when it is green. Only when the disk is green is the relation between pecking and food operative. In this fashion, an environmental stimulus that is not itself a reinforcer or a punisher exerts control over behavior by setting the occasion for the response-reinforcer relation, or letting the organism known that the response-reinforcer relation is in effect.

This kind of discriminative control over behavior by environmental stimuli is absolutely pervasive in human life. Very few human responses are reinforced in all circumstances. Different manners of dressing, of talking, of swinging a tennis racket, of writing, will be appropriate (reinforced) in different settings. Much of what is involved in building up an effective repertoire of actions is not so much learning how to do different things as it is learning under what circumstances the different things people already know how to do will be reinforced. The discriminative function of environmental stimuli is thus a crucial factor to be added to the relation between response and reinforcer that is specified by the principle of reinforcement.

Consider, now, the principle of reinforcement itself, the selection of responses by their consequences. First, what are to count as responses? An obvious answer is that responses are detectable movements of the organism. But this obvious answer won't do. First, there are lots of movements that should not be treated as responses. Parts of people move when they breathe, for example. Muscles twitch when people sleep. Glands secrete hormones when people are excited, or nervous, or sexually aroused. Although these various "responses" may well be important, they are not, with rare exceptions, the proper province of the principle of reinforcement. The kinds of responses one is typically interested in involve what are usually called *voluntary, skeletal* movements.

But restricting "responses" to voluntary skeletal movements is not good enough. Think for a moment about the pigeon pecking the disk. These pecks can be treated as voluntary, skeletal responses. It is reasonable to suppose, though, that no two of the pigeon's pecks are exactly alike. Some are faster than others, some last longer than others, some are more forceful than others. How, then, can one say that reinforcement increases the frequency of *pecking,* when each peck is unique? What justifies lumping all of these unique activities together? And if they can't be lumped together, then what sense does the principle of reinforcement make? It can't be that food is affecting the actual peck that preceded it. That peck is gone forever. So reinforcement of that peck must be affecting *future* pecks. To do this, however, requires that the pigeon's various pecks can somehow be treated as members of a single category.

Indeed, behavior theory treats responses not as unique entities but as members of classes. Each class is called an *operant.* This word is used because what characterizes the class is the way in which its members (individual responses) operate on the environment. An operant class may be defined very narrowly or very broadly. Thus, for example, "throwing" may be an operant, but so may "taking back the arm at a given velocity, releasing the ball at a given moment in the movement of the arm, and carrying the motion of the arm through until it reaches the knee." Or the global operant "throwing" may be broken down into "throwing footballs," "throwing baseballs," and "throwing darts," or into "throwing line drives," "throwing lobs," and "throwing grounders." There are no priori rules about how global or narrow an operant class will be. What determines its scope is environmental contingencies.

When a child is first being taught to throw a ball, any movement that propels the ball in roughly the right direction counts as throwing and is duly rewarded with praise from the teacher. As the child progresses, the teacher's standards tighten up. Only certain throws are acceptable and thus praised. In similar fashion, when the pigeon was being taught to peck the disk, the operant started out as "any orientation and movement in the general direction of the disk" and was refined so that only physical contact with the disk was acceptable. It could have been refined still more, so that the force and duration of each peck had to be within particular bounds. Through this process of redefinition of the operant by rearrangement of the environmental contingencies, the operant can become increasingly refined as the contingencies of reinforcement demand increasing refinement. What unites the members of an operant class is their environmental function. And what determines their function is the operative contingency of reinforcement.

Because operants are functionally defined, the actual form of the responses that comprise them may be highly varied. The pigeon might "peck" the disk by striking it with its wing or even by standing on its head and striking it with a foot. Each of these responses would count as members of the operant class "disk pecks," although they have virtually nothing physically in common with a peck. They are functionally equivalent because the environmental reinforcement contingency treats them as equivalent. The feeder doesn't care how the disk is struck. Similarly, environmental contingencies that operate on people demand that they get to work on time. Whether they do this by driving, taking a train, flying, running, or camping outside the office is of no consequence; the boss doesn't care how they get there, as long as they do.

The reason for discussing the definition of the operant in such detail is that it is a feature of the behavior theorist's scheme that captures especially well the flexibility of human behavior. Human actions do not come prepackaged into units; they get shaped or carved into units as a result of experience. A man goes to work making belts in a factory. How does he do the job? He might do it the way the guy next to him is doing it. Or he might do it the way his father before him did it when he made belts. Or he might do it the way his contract stipulates it should be done, or the way the foreman insists that it be done. Or he might do it lots of different ways, discovering by trial and error which method is most efficient and effective. If he makes more belts, he makes more money. In this way, the contingency of reinforcement that is operative will shape effective belt-making behavior. In this way, according to behavior theory, all operants get shaped.

Many people who know only a little bit about behavior theory fail to appreciate how flexible the concept of the operant is. They think of behavior theory as an attempt to explain the richness and complexity of human behavior as a collection of little muscle twitches, each triggered by some environmental stimulus. And because an explanation in these terms is so implausible, they dismiss behavior theory out of hand. It should be clear that this is a serious mistake. Operants *can* be muscle twitches, but they can also be as large and complex as "going to parties," "reading the newspaper," "cooking," "working," and so on. It is the dictates of the environment, not of the behavior theorist, that determine just how large and complex an operant will be.

What, then, of the second major category involved in the principle of reinforcement? What kinds of environmental events count as rewards and punishments? Here is a major parallel between behavior theory and economics. Economics makes no distinction between needs and wants; it has no theory of value. Individual desires will be particular and idiosyncratic, and all one can say is that people will always want something and will act in such a way as to maximize want satisfaction. So it is with rewards and punishments. Behavior theorists spent many years trying to identify some property that all rewards had in common—trying to develop a theory of value based on the biological needs of an organism. All such attempts failed. Sure there is a universal need for food and water, and food and water seem to work as reinforcers for all species, but plenty of other things also work as reinforcers—for some species but not others.

With people, there is even variation from individual to individual in reinforcer effectiveness. There is also variation in a single individual over different periods of time. Some environmental events that were once of no consequence can become reinforcers. Some can even go from being noxious to being reinforcing (for example, alcohol is an

acquired taste). In the final analysis, all that can be said about reinforcers (or punishers) in general is that they are all reinforcing (or
punishing). That is, reinforcers *increase* the future chances that
responses that produced them will occur, while punishers *decrease*
the future chances that responses that produced them will occur. An
organism's wants are individual and idiosyncratic, just as the economists say. And responses that bring satisfaction of wants are reinforced. Furthermore, behavior theorists would agree with economists
that organisms always want something—that wants are unlimited.
After all, all operant behavior is controlled by reinforcement, and
organisms are virtually always engaged in one or another class of operant
behavior. Thus, it is fair to say that economists and behavior theorists
alike see life as the constant pursuit of the satisfaction of wants.

Now that we have discussed the concepts of operant and reinforcer, let us turn to the relation between them. This relation is
easiest to describe by reference to a concrete example. Consider a
man who works from nine to five as an automobile mechanic, then
goes home to pursue his hobby—restoration of old cars to running
order. On his job, fixing cars is an operant. The weekly paycheck is
the reinforcer. He would not be fixing cars were it not for the paycheck, and he would just as soon do some other kind of work for an
equivalent or greater paycheck. Thus his job, the operant, is a means,
and his paycheck, the reinforcer, is an end, and there is no special
relation between the means and the end that could not be duplicated
by substituting some other job or operant for his current one.

The situation is different when he gets home. Now, fixing cars is
both means and end, operant and reinforcer. While it is true that he
does not tinker with cars just for the sake of tinkering—achieving the
goal of a smooth-functioning automobile is an important influence on
his activity—it is also true that he wouldn't be satisfied with any old
means to achieving that goal. He would not, for example, be satisfied
with hiring someone else to restore the old cars for him. The reinforcing consequences of the activity are a part of the activity itself,
and the other kinds of activity are not interchangeable with it in the
service of the same reinforcer. Indeed, we might even say that "owning old cars that run well" is not properly even a reinforcer, for it
will not increase the likelihood of any operant except for "fixing old
cars." Similarly, the operant of "fixing old cars" is not properly an
operant, since it will not be reinforced by any reinforcer except for
"having old cars that run well."

The distinction between this man's job and his hobby is one that should be familiar to all of us. Some people have jobs that are like this man's; they are simple means to an end—pure operants performed solely for the wage (reinforcement), which would be given up immediately if an opportunity for a bigger wage came along. Other people are fortunate enough to have jobs that are more like this man's hobby. While the wage is certainly significant and without it people wouldn't do the job (just as for the hobbyist, having a finished, working automobile is crucial and without it he would abandon his hobby), it isn't everything. There are aspects of the activities involved in the job itself that make it more than just a means and make people unwilling to substitute other jobs that pay just as well or better. So even though people work at these jobs for the wage, the job themselves are both operant and reinforcer. If there were no wage, people might behave like the car mechanic, taking jobs that paid and performing this activity as a hobby.

It is essential to behavior theory principles that the relation between operants and reinforcers be like the hypothetical man's job and not his hobby. The goal of behavior theory is to reveal something about the relation between operants *in general* and reinforcers *in general*. The principle of reinforcement indicates how any operant behavior whatsoever will be influenced by any reinforcer whatsoever. No one, after all, is especially interested in finding out all there is to know about pigeons pecking at disks for food. For this kind of generality to be achieved, the behavior theorist must be able to substitute operants for one another or reinforcers for one another, without substantially altering the way in which the latter affects the former. The man's nine-to-five job satisfies this requirement admirably. Since he is working only for the money, it is arbitrary, or accidental that his work happens to be fixing cars. It could just as well be building houses. And because this work–pay relation is arbitrary, what one learns from it can be applied to all such work–pay relations. This generality of application does not apply to his hobby, however. Other operants cannot be substituted for car fixing, or other reinforcers for having fixed cars. If the man's hobby were stamp collecting, or chess playing, the operant, the reinforcer, and the relation between them would all be different than they are for old car restoring.

It is assumed that when a pigeon is trained to peck a disk for food, an arbitrary operant–reinforcer relation is being established. It is assumed that the pigeon wouldn't peck disks as a hobby and that there is no special relation between pecking as an operant and food as a reinforcer such that if either were changed the principles discovered would be materially affected. These assumptions are what give the behavior theorist confidence that much can be discovered about the determinants of behavior *in general,* across different species and different situations, by studying a simple pigeon, in a simple box, engaging in a simple response, for a simple reinforcer.

Once we appreciate that the principles of behavior theory are restricted to operants and reinforcers whose sole relation is the contingency imposed by the environment, the potential generality of the principle of reinforcement is enormous. People will surely differ from one another, across history and across cultures, in the kinds of operants they perform and the kinds of reinforcers they obtain as a result. But the principle of reinforcement will apply to all sets of properly independent operants and reinforcers. In fact, there is no reason to restrict its application to people; it will apply as well throughout the animal kingdom. Different species are capable of different kinds of behavior. People can't fly and birds can't type. They will be interested in different types of reinforcers. People don't eat worms and birds don't eat chocolate. But across all species, historical periods, and environmental contexts, the principle of reinforcement will still hold—any operant of which an organism is capable will be increased in likelihood by the occurrence of any reinforcer in which the organism is interested.

It is behavior theory's confidence in the enormous scope of applicability of the principle of reinforcement that allows behavior theorists to derive most of what they claim to know about behavior in general from the study of a few very simple situations—situations like the pigeon pecking a disk, or a rat pressing down on a small metal lever, or running through a maze, to obtain bits of food, or to escape painful electric shock. The animals perform these tasks in highly restricted, simplified environments in which virtually nothing aside from operant and reinforcer ever occurs. They are environments stripped to their essentials, in which the principle determinants of behavior are laid bare. They are environments in which the principle of reinforcement operates like a precise machine producing effects on behavior that are perfectly lawful and predictable. They are environments which, although simple to the point of being impoverished, reveal the key to understanding human nature amid all the complexity of real life. That, at least, is what the behavior theorist claims.

The Economics of Behavior

Armed with this general explanatory framework and these central concepts, what does the behavior theorist do? What does he find out about the ways in which the principle of reinforcement influences operant behavior? If all behavior theory had to offer us was the principle of reinforcement, there would be little reason to pay much attention. After all, everyone has always known that people's actions are affected by their past and anticipated consequences. What behavior theory has to offer is a body of experimental results that detail how manipulation of various aspects of reinforcement influences behavior. Studying pigeons pecking disks and rats pressing levers, for food or for water, behavior theorists ask questions like these: (1) How does the quality and quantity of a reinforcer affect the likelihood of an operant? (2) How does the proximity in time of operant and reinforcer affect the likelihood of the operant? (3) What happens to the operant if it does not produce reinforcement every time it occurs? Can it still be sustained?

Studies of these questions have revealed a number of general principles. In general, the greater the amount or quality of reinforcement available, the more operant responding occurs (Fifty thousand dollars a year is a better reinforcer than fifteen thousand dollars). The more immediately reinforcement follows an operant, the more operant responding occurs ("nice shot" is more effective on the court than it is in the locker room). Indeed, in discussing cigarette smoking before, we saw the importance of immediacy of reinforcement. If the immediate consequence of cigarette smoking is positive, this can outweigh the long-term consequence, which may be *extremely* negative. Also, as in the pigeon example before, operant responding can be maintained even if reinforcement does not follow each and every response; reinforcement can sustain behavior when it occurs intermittently. And with suitable training prcedures, a little reinforcement can go a long way. One can get rats to produce more than a thousand presses on a lever for each reinforcement. Monkeys will work steadily for eight hours at a time if their day's ration of food is produced at the end.

One might look at the principle of reinforcement as it operates in the animal laboratory and be duly impressed by its power to control behavior. But one might skeptically ask, "So what? What does research with animals in cages have to do with real life? What does it have to do even with the real life of animals in nature, let alone of people in their complex everyday world?" The behavior theorists' response to this skepticism would appeal to the firmly established experimental tradition of science. Of course the pigeon's experimental environment is highly simplified and artificial. But so is the chemist's test tube, the physicist's vacuum, the biologist's cell culture. It is the point of experimentation to simplify and control what in nature is complex and uncontrollable. Once fundamental principles have been identified in the laboratory, the scientist makes good on her presumption that they operate outside it by putting them to work. The principles are applied in real life, and in that way it is shown that they are not peculiar to test tubes, vacuums, cell cultures, or rats and pigeons in cages.

Behavior theory can certainly claim to have made good on this score. Its principles have been widely and successfully applied to the behavior of animals and to the behavior of people. Behavior theory principles have proven effective in the treatment of a wide range of behavior problems that bring people to therapists or mental hospitals. They have been used effectively to treat alcohol abuse and to establish programs of weight control. They have been successful in eliminating disruptive behavior in the classroom and in establishing techniques that allow children to learn rapidly and efficiently. They have been used successfully for many years in the workplace.

One might be tempted to write these successful applications off as obvious and unimpressive. People have always known that bribes work, and that, after all, is what the principle of reinforcement says. But behavior theory also indicates how they work, when they work, and under what conditions they work most effectively. The result has been a technology for the modification and control of behavior that is most impressive and that could not have been totally obvious since the technology did not exist until research in behavior theory established it.

There is still room for doubt about the relevance of behavior theory principles, even in the face of successful applications. The ubiquitous feature of human life is choice. The issue for people is not whether to engage in some operant for reinforcement. The issue is *which* operant, for *which* reinforcement. When the economist discusses "rational economic man," he is discussing how people *choose* to allocate limited resources among available alternatives. He is discussing how people choose among different bundles of commodities so as to maximize preference. Choice is what economic behavior is about.

While it is true that even a rat pressing a lever has a choice available (to press or not to press), this is surely a degenerate form of choice. It is degenerate because it is hard to imagine what could possibly influence the rat not to press the lever in a setting in which few alternative actions and no alternative reinforcers are available. So while behavior theory may tell us about human behavior in the specific, impoverished conditions in which no real alternative courses of action are available, such conditions have almost nothing to do with everyday life as people know it.

In reply, the behavior theorist can say that he does have something to offer about the determinants of choice. He can give the pigeon, the rat, or the person, for that matter, a number of alternative operants to perform, for a number of alternative reinforcers, and see how the animal allocates its resources. Indeed, it is here, in the study of choice, that behavior theory and economics meet most clearly and explicitly. In the last decade, behavior theorists have borrowed the formal, mathematical framework of microeconomic theory and put it to the test, in the animal laboratory, asking in effect whether pigeons and rats are also "rational economic men." And the answer, in a word, is that they are.

Let us look at how such studies are conducted. In economics, the limited source that is being allocated is typically money, and for pigeons or rats, money is in notoriously short supply. So the resource behavior theorists study is behavior. How do animals allocate their time and energy when these can be devoted to a number of different sources of reinforcement? How will choice be affected by the quantity of the reinforcer available, the quality of the reinforcer, the delay of the reinforcer, and its cost (cost measured again not in money but in time or responses required to produce it)?

Behavior theorists ask questions like these by giving pigeons *two* disks to peck at. The reinforcement associated with pecks at either disk can be varied in type, amount, delay, or cost. Suppose, for example, that by pecking at disk A, the pigion will earn four seconds of access to food for every twenty-five pecks, while for pecking at disk B, they pigeon will earn four seconds of food for every fifty pecks. It doesn't take the pigeon very long to do exactly what a person would do: buy the food where it is cheaper. The pigeon quickly comes to make all its pecks at disk A. From here, one could now find out about the pigeon's scale of value for access to food. One could increase the amount of feeding time available for pecks on disk B—to five seconds, six, seven, and so on, until the point was found at which the pigeon's preference switched from A to B. We might expect the switch to come when feeding time on B exceeded eight seconds, since eight seconds is where the unit cost (seconds of food per response) on the two disks is equal. But perhaps, psychologically, eight seconds of food is less than twice as much as four seconds. Perhaps pigeons, like people, evidence diminishing marginal utility so that each little increase in the amount of food offered the pigeon is worth a little less than the one before it. If that were true, we might have to provide considerably more than eight seconds of food on disk B before the pigeon switched its preference.

The particular economic ideas that have been tested in studies of choice in animals come principally from a part of economics known as consumer demand theory. Consider a few examples of the testing and confirmation of consumer demand theory in the animal laboratory.

DEMAND ELASTICITY

Among the best-known principles of economics is the "iron law of supply and demand" and how the two combine to influence price. When supply exceeds demand, prices drop, and when demand exceeds supply, prices rise. The marketplace achieves a state of equilibrium with respect to a particular commodity when the amount of it that people are willing to buy at a given price equals the amount that suppliers are willing to make and sell at that price. But price changes do not affect demand for all different types of commodities in the same way. With some commodities—what might be called necessities—demand will stay high even if prices rise. People will purchase bread, milk, housing, medical care, and things like them no matter how much they cost. Demand for items like these is said to be relatively *inelastic,* that is, relatively insensitive to changes in price. In contrast, with other commodities—steaks, pastries, movies, and so on—demand will be dramatically affected by price. Demand for items like these is said to be relatively *elastic.* Certain commodities are relatively elastic for some people and inelastic for others. When the price of gasoline and heating oil tripled a decade ago, some people started driving less and keeping their houses colder. Other people went on consuming as before.

Differences in demand elasticity can be demonstrated in animals. Do they too distinguish luxuries from necessities? The answer is yes. In one experimental demonstration of the importance of demand elasticity, rats were given the opportunity to press two levers. Presses on one of them produced food. Presses on the other produced bursts of electrical stimulation of the rat's brain in an area that is sometimes known as a "pleasure center," so called because rats, and other animals, will engage in operant behavior to the point of exhaustion to produce such stimulation. When the cost of either food or brain stimulation was two lever presses, rats vastly preferred the brain stimulation to the food. From this one might be tempted to conclude that brain stimulation was much more *valuable* to the rat than food was. But when the cost of each was increased to eight lever presses, the rats' preferences changed. Now, they preferred food to brain stimulation. The specific difference between the two situations was not in the rats' consumption of food. They worked for almost identical amounts of food in both cases (demand for food was inelastic, that is, unaffected by its price). However, when brain stimulation increased in price from two to eight presses, consumption dropped to about one-tenth its previous level (demand for brain stimulation was elastic, that is, affected by price). So differences in demand elasticity make a difference in how a rat allocates its resources when prices change.

DEMAND AND INCOME

Viewing changes in demand as a function of changes in price gives an incomplete picture of economic behavior. What, after all, is price? It seems obvious. Price is what something costs—$1.98 a pound; $.79 a quart. But these dollar amounts are meaningless unless it is known how much a dollar is worth. Saying that demand is affected by price requires knowing something about prices in relation to income.

What, for a pigeon or a rat, is income? Since the price of things is measured in responses, income must be measured in responses as well. Traditionally, behavior theory experiments are conducted for fixed periods of time, say for a one-hour session every day. Within that time, the animal is free to make as many responses as it wants. Indeed, since one of the behavior theorist's main interests is in how various contingencies of reinforcement will affect the amount of responding an animal does, the animal's responding is virtually always left free to vary. From an economic point of view, however, this is equivalent to allowing the animal to print money. As the cost of food or whatever reinforcer is available goes up, the animal can always increase its income by responding more than it did before. Under conditions like these, the effects of price changes might well be dampened. To allow the effects of price on demand to show their true colors, one has to be able to change prices *relative* to income. To do this, one must control the income that an animal has.

A few experiments like this have now been done. Animals are given a fixed number of responses that they can "spend" during an experimental session. The session ends when they use up their income. Under conditions like these, highly elastic demand for certain commodities shows up very clearly. For example, in one such study, baboons could respond either for food or for heroin (infused into the skin through a surgically implanted tube). When income was under their control (that is, they could respond as much as they wanted), neither food choices nor heroin choices were very much affected by price (required responses). Thus, demand for both seemed inelastic. But when an income constraint was imposed so that the baboons could only make a certain number of responses per session, things changed. Now, demand for food remained relatively insensitive to its price, while demand for heroin was markedly affected by price. That is, heroin turned out to be a luxury once the animal's income was fixed. So it is true for animals, as for people, that the effects of price changes on consumption have a lot to do with the relation between prices and income.

DEMAND AND SUBSTITUTION

This discussion of demand has come a little closer to reality with the inclusion of effects of income, but there is still plenty missing. Demand for necessities is relatively inelastic and demand for luxuries is relatively elastic. But what makes something a necessity? Food is a necessity; people need food to live. True enough, but they don't need bread or milk. If the prices of those went up, people could substitute rice and cheese. If the price of beef goes up, people substitute pork or poultry. The key word here is *substitution*. Part of what determines the elasticity of demand for a commodity is the availability of appropriate substitutes. If substitutes are available, demand may be extremely elastic, even for apparent necessities. With margarine around, there are limits to how much people will pay for butter. Margarine is not a perfect substitute; butter tastes better, which is presumably why people are willing to pay more for it. But they will only pay a certain amount more. The importance of substitutes is clear when we think about the dramatic petroleum price increases of a decade ago. People *want* gas so they can transport themselves by car; car travel is very convenient. But there are substitutes. People can and will use public transport if the price of gas gets high enough. In places where public transport is inconvenient, as in much of southern California, demand for gas may be less elastic. But even here, there are substitutes—in the form of fuel-efficient small cars to replace gas-guzzling monsters. In the case of heating oil, people can keep their houses colder. They can also invest in substitutes, by converting to natural gas, or by installing passive solar systems, or by adding insulation. In a word, demand elasticity is critically dependent on whether alternatives are available.

The importance of substitutes is as true of the pigeon as it is of the person. Go back for a minute to the hypothetical experiment described before in which a pigeon's pecks on disk A produce four seconds of food for every twenty-five pecks, while pecks on disk B produce four seconds of food for every fifty pecks. Recall that the pigeon in this case spends all of its time pecking disk A. The reason for this is that the commodities available for pecks on disks A and B are perfectly substitutable—they are the same commodity. A tiny price advantage will be enough to produce a complete preference for the cheaper food. But suppose instead it was four seconds of water, rather than food, that was available on disk B. Now we would see a very different pattern of behavior. The pigeon would peck both disks, "purchasing" requisite amounts of food and water. Not only that, but if the price of food decreased—to say ten responses, this would result in an *increase* in the amount of water the pigeon purchased. That's right; even though the cost of food has gone from half the cost of water to a fifth of the cost of water, the result of the change is to increase water consumption. This is because not only are food and water not substitutable for each other, they are actually the opposite of substitutable; they are *complements*. The more food the pigeon has, the more water it wants. So as the price of food decreases, the pigeon's demand for water increases.

INDIFFERENCE CONTOURS AND BUDGET CONSTRAINTS

In real life, people obtain a fixed income (so many dollars per week) and have to decide how to spend it. A person earning two hundred dollars a week must decide how much to spend on housing, food, clothing, entertainment, and so on. This situation is one involving choices among different bundles of commodities, with the bundles differing from each other in how much of each commodity they contain. As an example, imagine going to a bakery to buy bread and cake. The money can be used to purchase rolls, or pastries, or some combination of both. If each roll or pastry cost twenty-five cents, one could purchase a wide range of "bundles" for a given amount of money. What kinds of bundles would people prefer?

Suppose that there was no significant distinction to be made between rolls and pastries. This would mean that rolls and pastries were completely substitutable for each other and people wouldn't care how much of either they had. So twenty rolls and no pastries, twenty pastries and no rolls, ten of each, or any other combination totaling twenty would be equivalent. People would be described as indifferent among these various bundles. However, they would prefer any of them to a bundle containing, say, ten pastries and eight rolls. It is unlikely, of course, that rolls and pastries are completely substitutable. One might be willing to trade sheer quantity for a little variety, preferring, for example, eight rolls and seven pastries to twenty of either. That people would almost surely prefer some variety to all of one or the other commodity is another face to the principle of diminishing marginal utility that we already encountered; the twentieth roll is worth less than the nineteenth, which is worth less than the eighteenth, and so on.

Economists assume that there are many different bundles of commodities among which people are indifferent, and all of them together are called an *indifference contour*. They further assume that each of us has many different indifference contours and that people prefer any of the bundles on some of them to any of the bundles on others. Finally, they assume that when people actually choose a bundle, they choose one on their highest possible indifference contour. This is an important part of the *rational* in "rational economic man." But what is the highest possible indifference contour that people supposedly choose? I would probably prefer a hundred rolls and a hundred pastries to a lesser number of either. Are there any constraints on my preferences?

On preferences, perhaps not, but on choices, certainly. The constraints are imposed by one's budget. If I have only five dollars to spend at the bakery, it doesn't matter that I prefer fourteen rolls and seven pastries to twenty of either. If each item costs a quarter, I simply can't buy more than twenty. So when people actually choose, they choose a bundle that is on their most preferred indifference contour that can be satisfied within the existing budget constraint.

In the case of people choosing among commodity bundles, all of this business about indifference contours and maximizing preference within budget constraints is assumed. It could hardly be otherwise; we can't sit people down and have them tell us their rankings of thousands of different bundles of commodities so that we can actually get a depiction of their indifference contours—what is called an indifference map. But with animals, this indifference analysis can be tested empirically. The details of the tests are too complex to discuss; they involve studying how preferences for commodity bundles change as the prices of each of the commodities involved change and as the budget constraint changes. The outcomes of such tests provide strong support for the economist's analysis. So even though one can not be sure that "rational economic people" choose bundles from the most preferred indifference contour within the budget constraint, it is clear that rational economic rats and pigeons do.

This indifference analysis takes its most general form if we break the universe up into bundles containing only two commodities—income and leisure. Income can be used, of course, to purchase all sorts of commodities; leisure purchases only itself. If organisms must work for their income, then there is direct trade-off between the two different commodities comprising a bundle. More income means more labor and less leisure; and conversely, more leisure means less income. Now what bundle of these commodities does one prefer? Economists generally assume that bliss is all leisure and no work. Behavior theorists, in defining the operant the way they do, make the same assumption (remember the distinction between fixing cars as an operant and fixing cars as a hobby; the critical thing about operants is that people do them *only* for the reinforcer). Real life allows very few this state of bliss, since without work people would have nothing to do during their perpetual leisure but starve.

Still, given that people have to work, there are many questions that economists ask about how trade-offs between labor and leisure are affected by such things as rates of pay, or the availability of welfare payments, or a host of other things. A major line of criticism of social transfer payment schemes such as welfare or negative income tax is that they destroy people's incentives to work. But despite the heat that has been generated in disputes about the effects of welfare programs on willingness to work, it has proven very difficult to shed empirical light on the issue, principally because there are so many other, uncontrolled factors operating in real human societies. Indeed, it has proven very difficult even to figure out whether increases in wages increase or decrease the amount of time people spend working. On the one hand, one would expect that if someone's salary were raised from two hundred to four hundred dollars a week, he would start choosing labor-leisure bundles with more leisure in them, since he can get the same income as before with half as much work. But on the other hand, this pay raise has also increased dramatically the price of leisure for this person—in potential wages foregone. If this price effect is big enough, wage increases will produce more rather than less work; if not it will produce less rather than more.

Again, while with people we largely assume or argue about the labor leisure indifference map, with animals in the laboratory it can be studied—and it has been. Pigeons will sacrifice some food for leisure, how much depending on both their income and the price of food. Some kinds of welfare schemes will interfere with the pigeon's incentive to work, but others will not. The behavior theorist's laboratory has become a place where the economist's indifference analysis can be subjected to careful experimental evaluation.

SPENDING AND SAVING

Economic agents, choosing among bundles of commodities, are often faced with another kind of choice. This is the choice between purchasing any one of the infinite possible bundles of commodities and choosing none. That is, people have a choice between spending and saving. More accurately, people have choices to make about *how much* to spend and *how much* to save.

If people spend their money right away, they can enjoy the benefits attached to the commodity bundle they purchase right away; satisfaction is immediate. If they save, satisfaction will be deferred. So the choice between spending and saving is a choice between immediate satisfaction and delayed satisfaction. (Note, of course, that this way of looking at things does not apply to people whose income is just enough for subsistence; these people don't really have a choice to make between spending and saving.) How does one choose when confronted with a choice between immediate satisfaction and delayed satisfaction?

One thing is clear; if someone is choosing between a given satisfaction now and the very same one later, it is hard to imagine why he would put it off. That is, there seems to be no good reason to delay gratification just for the sake of delaying it. An ice cream cone now ought to be preferred to the same one three hours later. Another way of saying the same thing is that waiting time discounts the value of commodities. Buying a ticket for tommorow's show is worth less than buying a ticket for today's at the point at which the ticket buying is done. Given this fact—what economists usually call *temporal discounting*—if people are going to defer purchasing of commodities, it has to be worth their while. One of the economic-social institutions that makes savings worthwhile is the interest people earn when they put money in the bank. Indeed, interest is a way of compensating people for deferring their gratification. We can even assess how steep a person's temporal discount function is by seeing just how much interest we have to provide to get that person to save. A huge rate of interest would be needed to get someone to save for whom the value of some commodity sinks very rapidly as its availability is deferred to the future.

We have already encountered the temporal discounting of commodities by animals in discussing the importance of the immediacy of reinforcement. Food that follows a response immediately is much more effective than food that follows the response after some delay. Also, in discussing cigarette smoking, we saw how relatively trivial positive consequences can dominate rather substantial negative ones if the positive consequences are immediate while the negative ones are delayed. For people, as well as for pigeons and rats, a bird in the hand seems to be of critical importance.

What kind of temporal discounting is consistent with "economic rationality"? This, it turns out, is an impossible question to answer. How much it is rational to save will depend on what one is saving for, how much the deferred commodity means to you in comparison with immediately available commodities, and other factors like these. But since there is no way of knowing how much that future commodity means to someone except by looking at what he does now to get it, virtually any pattern of saving is rational. If one saves a lot, it must mean that he wants that deferred commodity a lot; if one saves little, it must mean that he doesn't want that deferred commodity so much. Either way, he is being rational.

So in general, we are at a loss to call some patterns of saving and spending irrational and others rational. But there are plenty of times when the people involved help out. They tell us that they want to save (for a vacation, a car, a house, college for their kids) but that money burns a hole in their pocket. If they have it, they spend it. They can't resist the temptations of immediate consumption. These people, we can say, are being irrational. They are acting against their own interests *as they describe them*. How do we help them overcome this irrationality, this impulsiveness?

One thing we do is set up institutions that essentially take the decision to spend or save out of their hands. Mandatory retirement pension plans are an example. More interestingly, sometimes we get them to commit themselves to saving at a time when the temptation to spend is not so great. Payroll deduction savings plans are an example of this. On Monday, with payday far away, Jane signs a paper authorizing her employer to deduct twenty dollars a week from her salary and place it in a savings account. She never would have forked over the twenty dollars herself on payday, but she is willing to do it on Monday. How come? The answer is that on Monday, no consumption can be immediate. While consumption from savings may be very far off, consumption from the paycheck is also delayed. So now, the choice is not between immediate and delayed consumption but between two delayed consumptions. Deferring gratification by saving is easier under these conditions.

What makes this example interesting is that the person who is forcing the savings on Jane is Jane herself, the very same person who would choose not to save if the money were in her hand. We might be inclined to say that Jane knows what is good for her but also knows she is too weak to do it. So she binds or commits herself to doing the right thing at a time when the temptation is not so strong because she knows that she won't be able to resist the temptation if it gets any stronger. A very famous example of this sort of self-imposed restriction to resist future temptation comes from Homer's *Odyssey*. Here, Ulysses orders his crew on board ship to strap him against the mast. He knows that they will be passing near the island of the sirens and that he will not be able to resist their seductions, though he should. So Ulysses binds himself (literally, to the mast) at a time when the seductions are absent so that he will stay the course later when their presence is too powerful to resist. Note how paradoxical this self-imposed restraint is. Ulysses knows that he will want to visit the sirens when his ship draws near. If the business of life is want satisfaction, he ought to do whatever he can to make the satisfaction of that want possible. Yet, he does the opposite. This suggests that he really doesn't want to visit the sirens. But if that were true, he wouldn't have to strap himself to the mast in the first place. Strange business this economic rationality.

The reason for the foregoing discussion is that the phenomenon of temporal discounting, and possible techniques for delaying gratification, can and have been studied by behavior theorists. Pigeons also like immediate gratification, even when it is against their own best interests. Indeed, pigeons seem worse than people, at least worse than adults (there is a great deal of evidence that the ability to delay gratification increases with age; the temporal discount function gets less steep as we get older). If they are given a choice between a small amount of food immediately or a much larger amount of food a little bit later, they will go for the small, immediate payoff nearly every time. Yet, like people, if given the opportunity pigeons will bind themselves, committing themselves to the large, delayed payoff instead of the small immediate one at a point in time when the temptation of immediacy is less strong.

Armed with the finding that pigeons resemble people in the way in which they discount commodities over time and in their ability to opt for imposed restraint that allows them to overcome impulsiveness, behavior theorists have studied the determinants of impulse control in some detail. They have discovered techniques that strengthen control over impulsiveness by pigeons, some of which have been used successfully to instill "self-discipline" in people. As in the case of the other economic concepts we have discussed, the behavior theorist has been able to bring the economic concept into the animal laboratory and to test and refine it in ways that economists working with real-life human society can not.

RATIONAL ECONOMIC PIGEON: OPPORTUNIST OR STRATEGIST?

The foregoing examples of research conducted in the behavior theorist's laboratory were intended to show that the economist, studying the rational pursuit of self-interest, and the behavior theorist, studying the control of operant behavior by reinforcement, are really studying the same thing. Both disciplines are fundamentally concerned with the way in which organisms allocate their scarce resources so as to satisfy their wants. And both disciplines can make a case that the business of want satisfaction is affected in predictable, orderly ways by variations in factors like price, income, delay, and the availability of alternatives. The two disciplines seem to fit like a hand in a glove. Economists make *assumptions* about what the economic behavior of individuals is like and use these assumptions to predict the economic activity of groups—aggregations of economic individuals. But they rarely subject their assumptions about individuals to empirical test. The behavior theorist, in contrast, actually examines the economic behavior of individuals. And so far, it seems fair to say that the assumptions economists have been making are not far off the mark.

And yet, despite this happy alignment of behavior theory and economics, it is possible that the similarity between them is only superficial. Pigeons and rats may behave like "rational economic men," but the processes involved in controlling their behavior may be completely different from the processes involved in controlling human economic behavior. That the final products are the same does not mean that the processes giving rise to the products are also the same.

To see the sense behind this distinction between product and process, imagine two golfers, side by side, about to try to hit their balls to a putting green that lies one hundred and fifty yards in the distance. The first golfer strikes his ball and it flies straight and true, sailing in a high arc, and coming to rest on the green, just six feet from the hole. The second golfer misfires, and his ball starts to sail way off to the left of the green, into the woods. Then it hits a tree branch, and pops out of the woods, at a right angle to its original path, landing on the green, six feet from the hole. Same final product, but through rather different processes. If the second golfer is not mindful of the distinction between process and product, his future on the golf course will almost surely be miserable.

If in fact pigeon and person are producing the same final product via different processes, it may just be a happy accident of circumstance that the final products are the same. If the pigeon's behavior were studied under slightly different conditions (without a metaphorical tree branch in the woods to straighten the ball out), its resemblance to human behavior might disappear. And there is some reason to believe that the processes governing human and pigeon economic behavior may be very different, at least if we take seriously the differences in language used by economics and behavior theory in accounting for the phenomena in their respective domains.

The economist's conception of economic decision making is very much like our everyday conception of human action. The word *rational* in "rational economic man" is not just window dressing. The economist thinks of people as intelligent, as capable of formulating plans or goals that often reach far into the future, then of considering alternative courses of action that may achieve those goals, and finally of selecting the course of action that is most likely to be successful. People don't maximize preferences by accident or by magic; they calculate, evaluate, anticipate, and achieve preference maximization by design. The behavior theorist has no use for this kind of explanatory scheme. "Intelligent" action is not the result of planning and foresight; it is the result of selection of behavior that works by the principle of reinforcement. Well, suppose that the economist and the behavior theorist are *both* right, that people do it the economist's way while pigeons do it the behavior theorist's way. If this were true, the parallels between the two disciplines in final product would become a good deal less interesting and important.

To set the contrast between the economist's and the behavior theorist's stories, consider two different imaginary organisms: "strategic economic man" and "opportunistic economic man." The strategist goes through life like a chess master. He thinks many moves ahead in determining the consequences of his action, he considers a wide range of environmental possibilities corresponding to the different moves his opponent might make, and then, with full information about the possibilities, he chooses the move (action) that is to his maximal advantage. Acting in this strategic fashion may sometimes involve passing up a chance at a short-run gain to preserve the opportunity for a bigger gain later. So the strategist won't sell his house for a handsome profit now because he knows that a major corporation is about to relocate its headquarters in his town, which will almost surely drive the price of his house still higher. Indeed, the strategist may even make moves that are against his interests in the short run because they will serve them in the long run. Thus, not only does he not sell his house, but he spends ten thousand dollars building a tennis court (though he doesn't play) in the expectation that it will make his house worth forty thousand dollars more on the market. The strategist is truly out to maximize his self-interest in the long run.

Now what about the opportunist? He sort of stumbles about randomly, with no clear idea about long-range goals or about plans for achieving them. However, when he happens to encounter something he likes, he knows it. That is, he is able to tell when an action results in a short-term improvement in his self-interest. And when it does, he pursues it, and the long-term consequences be damned. He will sell his house for a profit, and he won't build a tennis court. While the strategist is on the prowl for the *best* alternative, the opportunist is only on the prowl for the *better* alternative. While the strategist *maximizes*, the opportunist *meliorates*.

Economists view organisms as strategic maximizers, and behavior theorists view them as opportunistic meliorators. Who's right? In fact, no one knows. In both behavior theory and economics, there are long-standing and lively disputes about which view is the correct one. Although economists by and large assume that organisms are strategic maximizers, some economists, most notably Nobel Prize winner Herbert Simon, have made persuasive arguments that the kinds of calculations that are required for real maximization are so complicated that they are either beyond human capacity or would take so long that the costs in time of doing them would outweigh the benefits. Think about the strategic house seller. Should he build a tennis court, or a swimming pool, or a patio, or an extra bedroom? Should he remodel his kitchen or hire a landscape architect? Should he really wait for the corporation to move in? Perhaps then everyone will be willing to sell their houses and the market will be glutted. The possibilities are endless.

Economists assume that people have complete information about the possibilities instantaneously and that gathering information in itself has no cost. In reality, of course, these assumptions are untenable. They are untenable for the individual consumer, and they are untenable for the business firm. The firm must decide what to produce, how to produce it, how much to produce, where to get materials used in production, where and how to market the product. The firm can certainly figure out that some courses of action will be better than others, but it would go out of business before it was finished calculating the one best course of action. What Simon suggests is that rather than maximizing, people and firms *satisfice*. They establish some criterion to determine what kind of outcome of production or exchange will be good enough rather than best, and when they find one course of action that satisfies the criterion, they pursue it. This is a lot closer to opportunism than it is to strategic maximization.

And most (but not all) behavior theorists move the organism another step closer to opportunism. Satisficing is not a deliberately selected decision-making procedure. There is no intentional setting of criteria for what is "good enough." All that is required is that organisms be able to recognize it when something comes along that is better than what they have already got and then shift their behavior in the direction of the better alternative. One is tempted to call this a kind of "evolutionary" view of economic decision making. Behavior evolves, slowly, in whatever direction is more effective at the moment, just as species evolve, slowly, in whatever direction is more effective at the moment. The major insight of modern evolutionary theory, recall, is that it is possible for well-adapted organisms to evolve without the application of *any* planning or intelligence—of even a satisficing sort— on the part of either the organisms' ancestors or of God. And so we come at last to another parallel between economics, behavior theory, and evolutionary biology. All three disciplines view organisms as moving in the direction of maximizing—of preference in economics, of reinforcement in behavior theory, and of inclusive fitness in biology. And in all three disciplines, maximization is the abstract ideal, unrealizable in practice because of limits on the organism's capacity to evaluate possibilities and to control environmental opportunities.

A Unified Science of Human Nature

We have reached the end of our discussion of some of the ways in which behavior theory has borrowed, and is studying, central concepts from economics. It seems clear that, although behavior theory started from quite a different framework than economics and still employs very different methods to study behavior and very different concepts with which to explain it, the two disciplines are converging. They are concerned with explaining the same thing—the allocation of scarce resources by individuals in such a way as to satisfy wants. It is not hard to imagine a time in the not too distant future when the various psychological processes that economists have simply *assumed* to characterize human economic activity will have been subjected to rigorous experimental scrutiny by behavior theorists. As a result of this scrutiny, some economic formulations may have to be abandoned, and others may have to be substantially revised. But the consequence of this disciplinary convergence may be a daunting unified science that uses a single set of principles to take us all the way from the economic behavior of single individuals—the atoms of society—to the economic behavior of massive macroeconomic systems.

One of the objectives of this book is to establish that evolutionary biology, economics, and behavior theory share a common vision of what it means to be a person, of what is essential about human nature. Having now completed a rather hasty tour through each of the disciplines, we are in a position to see what their common vision of human nature is.

In essence, human beings are economic beings. They are out to pursue self-interest, to satisfy wants, to maximize utility, or preference, or profit, or reinforcement, or reproductive fitness. They are greedy, insatiable in the pursuit of want satisfaction. As soon as one want is satisfied, another takes its place. If people ever stopped wanting they would stop doing altogether. Groups of people (societies) are just collections of individuals. The interests of society are the summed interests of its members. The wants of society are the agglomerated wants of its members. The good society is the successful society, and the successful society is the one that allows its individual members to satisfy their wants. The principle of natural selection operates at the level of the species in its natural history, of the individual in his life history, and of the culture in its social history to make people increasingly effective at satisfying their wants. Successful individuals survive and reproduce, proliferating their genes through the population.

Successful behaviors survive and reproduce, proliferating themselves through the life of the individual. And successful cultural practices survive and reproduce, proliferating themselves through successive generations, and spreading to other societies. This, in a nutshell, is what it means to be a person.

Now notice that while there is plenty about this picture of human nature that seems familiar, even right, there is also plenty that the picture apparently leaves out. This picture leaves out any consideration of morality, of how people *ought* to be as opposed to how they are. There is nothing accidental about this omission. We saw right from the start that it is the ethic of science to separate matters of morality from matters of fact and to concern itself only with the latter. And economists, biologists, and behavior theorists are all firmly committed to this ethic. Thus, one might want to argue that people ought not to be so self-interested, that the slavish pursuit of individual satisfaction is immoral, and the economist, biologist, or behavior theorist might *agree*. It is immoral, they might say, but it's the way things are. Denying it because one doesn't like it will only get people into trouble.

Indeed, they might continue, understanding this basic character of human nature is probably essential if cultures are to adopt codes of moral conduct that are effective. The real test of any set of social rules and sanctions is in the end not whether they satisfy some abstract, philosophical ideal of moral conduct, but whether they work. The real test of any economic system is not whether it satisfies principles of fairness, but whether it satisfies human wants. The real test of individual actions is not whether they conform to the Ten Commandments, but whether they promote the individual's survival. Society can establish social rules that attempt to suppress the individual's pursuit of self-interest, and they may even be successful for a time, but eventually essential human nature is going to burst through to the surface, or if it doesn't, a society whose rules permit the expression of basic human nature will arise to dominate and eventually replace this suppressive one. The best one can hope for is a set of moral rules that is constrained by the way people are—a set of oughts that conforms to the demands of the is's of human nature that these three disciplines are uncovering. It may be acceptable for an individual to decide that he'd rather be right than be president, but if whole cultures decide that they would rather be right than be productive, they are doomed to extinction. In the last analysis, what is right is what survives.

Thus, although the mere fact that these sciences of human nature ignore moral considerations does not mean that they deny their existence, it certainly does give them diminished significance. For the ultimate criterion for choosing among moral precepts is not itself a moral one; it is a natural one. Natural selection will operate on moral precepts just as it operates on everything else.

Now this kind of view raises an interesting apparent paradox. It is the ethic of science to keep its domain of inquiry morally neutral. What justifies this ethic? Do scientists behave this way because it would be immoral to behave otherwise and let morals intrude into scientific formulations? Would it be wrong to describe human nature the way people think it should be and take that description to be the way human nature is? If the ultimate criterion for moral neutrality in science is itself a moral one, then it looks as though the account of human nature offered by biologists, economists, and behavior theorists cannot be used to explain their own behavior as scientists.

Practitioners of these disciplines would claim that this paradox is only apparent. For what justifies moral neutrality in science is not any moral criterion but a natural one. Morally neutral science simply works better. It allows one to come up with a more accurate picture of the world than would be possible if people went out looking for what they wanted to be there. Moral neutrality, as an explicit part of the scientist's commitments, is something that has evolved in the history of science. And the reason that it has evolved is that it was better than its morally loaded predecessors. Sciences that see the world the way they think it "ought" to be as opposed to the way it really is simply fail to discover the truth. This, at any rate, is the way scientists justify the moral neutrality of science. What is right is what works best.

The convergence of these three disciplines is not entirely new. It represents the continuation and elaboration of a century-old tradition. Remember that a hundred years ago, defenders of the market system justified it by appeal to evolutionary theory. Unconstrained competition, which the fittest would survive, was nature's way. Indeed it was the only way to ensure continued social vigor. Just a few years before, Darwin had made his major conceptual breakthrough by analogizing competition for resources in nature to competition for goods in the marketplace. Two things have changed in the intervening century to make the convergence more formidable now than it was before. First, both economics and evolutionary biology have undergone considerable theoretical and empirical development without disturbing the parallels between them. This development makes each a more serious candidate for allegiance than it was before. Second, and more important, behavior theory has arisen, in the image of its forbears, to provide a crucial link between biology and economics. There is a big difference, after all, between evolutionary time as a scale of change and cultural time as a scale of change. There is a big difference between genes as media for information transfer and language and social custom as media for information transfer. There is a big difference between natural selection as a mechanism for determining behavior and rational calculation as a mechanism for determining behavior. Behavior theory bridges each of these gaps. It may be, for this reason, that behavior theory is the linchpin that is holding this triumvirate together. The pursuit of self-interest has been selected in evolution. It is selected in the life history of each individual organism. It is selected in the life history of a culture. Only cultural practices that cater to the pursuit of self-interest survive. The culture that caters best to the self-interest that is human nature will defeat all comers. Truly, this is the best of all possible worlds. Or is it?

The Limits of Economics

We cannot mediate all our responsibilities to others through prices. . . . [It is] essential in the running of society that we have what might be called "conscience," a feeling of responsibility for the effect of ones actions on others.

KENNETH ARROW

The outstanding discovery of recent historical and anthropological research is that man's economy, as a rule, is submerged in his social relationships. He does not act so as to safeguard his individual interest in the possession of material goods; he acts so as to safeguard his social standing, his social claims, his social assets.

KARL POLANYI

*C*hapter 3 painted a picture of human nature as seen through the eyes of a modern economist. To be more precise, it painted a picture of human *economic* nature, since economists disagree about just how much of human nature is economic. In that picture, rational economic agents, out always to maximize their self-interest and left alone to exchange in the free marketplace, create a prosperous, efficient society in which everyone gets what he wants, or, at least, in which more people get what they want than would be possible in any other form of social and economic arrangement. Now it is time to examine that picture critically, to determine just how accurate a portrait of human nature it is. What a critical examination is going to reveal is this:

1. The portrait of human nature provided by economics is, in some important respects, inaccurate. Some propositions about economic rationality are violated in significant circumstances; that is, people sometimes display a variety of economic "irrationalities." Other propositions about economic rationality are at best controversial.

2. The portrait of human nature provided by economics is importantly incomplete. Making sense of people's economic activities requires knowledge of their noneconomic ones, about the cultural institutions that influence their lives. Indeed, the smooth and efficient functioning of a market economy depends critically on aspects of people that are not themselves economic in nature.

3. How closely people approximate the economist's portrait of economic rationality depends upon the kind of culture they inhabit. Under some cultural conditions, the fit between the economist's portrait and human action may be quite close, while under others, it may not. This means that economic rationality as described by economists is less a reflection of an eternal, natural law than it is a reflection of particular cultural conditions that could be different if societies chose to make them so.

Economic Irrationality

According to the economist, people, as rational economic agents, can express preferences, prefer more of a given commodity to less, have transitive preferences, have relatively stable preferences, value commodities in decreasing amounts as they have more of them, and act with perfect information so as to maximize their preferences. It turns out that each of these characteristics of economic rationality is violated by people in important ways. In a word, economic rationality, as defined by economists, does not routinely exist in people. Let us consider some of the respects in which people are sometimes "irrational".

1. *People can't always express preferences among commodities.* A rational economic agent's ability to express preferences need not extend to everything. A person may find it impossible to say whether he prefers wealth, beauty, intelligence, or good humor in a prospective mate. However, within the domain of commodities, people should always be able to express preferences. Can they? Imagine someone with fifty dollars burning a hole in her pocket. Should she have a fine meal, buy a few shirts, go to the theater, or buy several records? After some

reflection, she may well be able to rank these options, which is to say that she can express preferences among them.

But these options do not exhaust the things that can be done with fifty dollars. It can be given to any of a number of charities, it can be used to buy groceries, it can be used to have the house cleaned, it can be used to buy school books, it can be used for part of the plane fare to a vacation spot, it can be used for part of the cost of having the house painted, or to have someone care for the lawn, or to look after the children. The list of things one could do with fifty dollars is almost endless. Can people express preferences among all of these different possibilities? Is a good meal preferred to having the house painted? Is child care preferred to a vacation? Is the theater preferred to charitable contribution? Everyone may be able intelligibly to express preferences among *some* of the things that can be done with fifty dollars, but no one can express preferences among *all* of the things that can be done with fifty dollars.

The reason for this inability to express preferences is that some sets of commodities are simply incomparable or incommensurable. There are no dimensions available for making the necessary comparisons. The only thing that all of these things have in common is that they cost fifty dollars. The satisfaction that that fifty dollars will produce is so different for different types of commodities that one is at a loss to know how to figure out what is preferred to what. The economist might reply to this concern that while people may *think* they are unable to express their preferences, they actually can. They might not have conscious access to their preference hierarchy, but they have one nonetheless. People do, after all, eventually do *something* with the fifty dollars. If they were really unable to express preferences, they would sit paralyzed with uncertainty as to what to do, while the fifty dollars accumulated interest in a bank account.

It is true that people do make choices, but it does not follow from this that they can express preferences among all the things they could logically be choosing from. The set of possibilities they actually consider may be only a tiny fraction of the set they might consider. Indeed, one way of thinking about just how people do go about making choices is that they organize the world of possibilities into a set of distinct categories. Categories might include such things as: household necessities, household maintenance, charity, one-night indulgences, longer-term indulgences, personal appearance, children. Now within each category it may be relatively easy to express preferences. A nice dinner might be preferred to the theater—at least tonight. Public broadcasting might be preferred to medical research as an object of charity. Painting the house might be preferred to fixing up the lawn. Between categories, however, expressing preferences is more problematic. On this view, when faced with the problem of spending fifty dollars, one must first decide what category of thing to spend it on. Once that is decided, one can follow the dictates of preference within a category.

This formulation raises several questions. How does one decide which categories to divide the world into? How does one decide which specific things go in which category? And how does one decide which category to devote this fifty dollars to? The economist has nothing to say about the first two questions. About the third he might argue that choosing a category is itself a reflection of preference. If one chooses a night out and not a charitable contribution, then one prefers a night out to a charitable contribution. Based on one understanding of preference, this is certainly true. That is the understanding of preference *as* choice. What is preferred is what is chosen, by definition. But the chooser may insist that she doesn't prefer a night out to charity—that she can't even compare them. Whatever the basis of her decision in this instance, it was not preference. She didn't even consider giving this fifty dollars to charity.

There are many noneconomic factors that might influence the way in which people categorize possibilities. Habit is one source of influence. Cultural norms are another. In our culture, clothing and hair care may both be considered as pertaining to matters of appearance. But one could easily imagine a culture in which what people wear has deep social, even religious significance, while how they keep their hair is a trivial detail. In that culture, a haircut and a new shirt would not be lumped together.

Is it important to economics that people may not be able to express preferences among all possibilities, that instead they express preferences within categories that are established by habit or social custom? Consider what this qualification on preference does to the notion of economic rationality. Can it be said that some ways of categorizing commodities are more rational than others? Do the canons of economic rationality apply to determining what goes with what? It seems they do not. It is no more or less sensible to lump haircuts and shirts together than to keep them separate. But if this is true, then knowing that a person is a rational economic agent reveals very little about his choices. It will not reveal which options he views (or should view) as comparable and which he views (or should view) as incomparable. It will not reveal which of his various incomparable categories of commodities will (or should be) getting his deliberation as he ponders what to do with the fifty dollars. All it can reveal is how he will choose from within a category given that he has already established the categories and established which one is worthy of consideration at the moment. And this is not very much to reveal. It leaves a good deal of the most important explanatory work to be done by factors that are not themselves economic.

2. *People don't always prefer the cheap to the expensive.* A rational economic agent will prefer six apples for one dollar to the identical six apples for two dollars. This is obvious, and so long as the choice is between identical commodities, paying the higher price would surely be irrational. But how often are people faced with this sort of decision? Are the apples in the two markets really identical? Maybe the ones that cost more are of higher quality. It takes a lot of work to find out just how good the various products one is choosing between really are.

One way of short-cutting is simply to assume that things that cost more are better. Consider trying to decide which video recorder to purchase from among six different brands, all with the same features. They range in price from, say three hundred to six hundred dollars. In response to a question about the price differences, the salesman provides a technical answer about the varying gradations of sophistication and quality of the electronic components contained in each of them. Now what? Does the salesman know what he is talking about? Does the customer? What is to be done?

One possibility is a study of electronics, to determine the technical merits of the different recorders. But by the time the study was complete, all these models would surely be obsolete and the study would have to start all over. A more reasonable approach is to assume that the more expensive a recorder is, the better it is, and reject the cheapest. And perhaps as a hedge against being played for a fool, also reject the most expensive. This leads the buyer to choose a mid-priced model from among the set of ostensibly equivalent commodities. How often, in the modern world, do people use price as a guide to quality, buying the more expensive of seemingly identical commodities precisely because it is more expensive? Is this irrational? If it is, then many people are frequently irrational; if it isn't, then price is not going to be a very good guide to preference. And note that this sort of price-guided purchasing is different from another kind of violation of the "cheaper is better" maxim that economists have known about for a long time. That is the so-called snob effect, that people will buy expensive things just to show that they can afford to buy them, to set themselves apart from others. This type of violation of "cheaper is better" is probably not irrational, since what people are purchasing here is not a commodity but evidence of status. Sociologist Thorstein Veblen, in his book *The Theory of the Leisure Class,* written at the turn of this century, gave such snob effects the label "conspicuous consumption."

3. *Preferences are not always transitive.* To say that preference is transitive is to say that if A is preferred to B, and B is preferred to C, it can be inferred that A is preferred to C. At first blush, it would appear that transitivity of preference is an essential element of economic rationality and one that everyone obviously possesses. If we like apples better than bananas, and bananas better than pineapples, we surely like apples better than pineapples. Or do we? Well, suppose a person has two different reasons for eating fruit: taste and nutritive value. Suppose further that given that the difference in nutritive value between two fruits is small, he will be guided only by taste, while if the difference in nutritive value is large, he will be guided only by that. Finally, suppose that pineapples have more nutritive value than bananas, which in turn have more nutritive value than apples, while taste preferences among the three fruits are exactly the reverse. Now, when faced with a choice between an apple and a banana, the consumer evaluates both nutrition and taste. A banana is somewhat more nutritious, but not a lot more. So he goes with taste and chooses the apple. Later, when faced with a choice between banana and pineapple he performs the same evaluation. Again, though pineapple is more nutritious, it is only a little more nutritious, so he goes with taste and takes the banana. Finally, he is faced with a choice between an apple and a pineapple. Transitivity says he will take the apple, but does he? While the difference in nutritive value between apple and banana may not be large enough to matter, and the difference in nutritive value between banana and pineapple may not be large enough to matter, the difference between apple and pineapple *is* large enough to matter. The criterion for choice flips from taste to nutrition, and he chooses the intransitive pineapple.

Does this example seem fanciful? Well, consider another, the purchase of a home computer. Suppose the choice has been narrowed to three candidates, and in deliberating, two things are relevant—price and available software. Candidate I costs four thousand dollars, and has lots of available software. Candidate B costs twenty-five hundred dollars, and has a fair amount of available software. Candidate M costs one thousand dollars, and has rather little available software, although it does have the essential programs. There may be many respects in which I is better than B and B is better than M, but in this example, only price and available software matter. Which is the rational choice? It is not at all implausible that a purchaser might reason roughly as follows: fifteen hundred dollars is not so much money over the life of a computer, while available software is crucial to determining how much use one will get out of it. So I is preferable to B, and B is preferable to M. Therefore, choose I. But wait. Not so fast. Three thousand dollars *is* a lot of money. If it is possible to get the essential software and save three thousand dollars, it's a good deal. So M is preferable to I. Intransitivity strikes again.

What, in general, can be said about intransitivities? When might they appear, and are they really intransitivities? In general, we can expect preferences to be transitive whenever there is only one dimension of comparison among commodities that is relevant. If all that matters to choice of fruit is taste, or all that matters to choice of computer is software availability, then if one is rational, preference will be transitive. However, whenever there are multiple dimensions that might be relevant to a decision, the potential for intransitivity is there. All that is required is that the relative importance of the various dimensions in influencing a decision can change, depending on the size of the difference between the candidates on each of the dimensions. Thus, a fifteen hundred dollar difference in price is not enough to let price determine a decision, while a three thousand dollar difference is.

The economist might object that these examples are not really examples of intransitivity of preference. If commodities are decomposed into the various dimensions that are relevant for comparison, rational people will show transitive preferences on each of these dimensions. That is, people will be transitive about taste, transitive about nutrition, transitive about price, and transitive about software. So then preferences really are transitive. Unfortunately, however, the objects that people purchase do not come decomposed into dimensions. They come as whole objects, and people choose *them* and the package of dimensions they contain. So even if preferences are transitive on every single dimension one can think of, it does not follow that preferences among actual objects will be transitive.

To complicate the matter further, relevant dimensions are not written in stone, eternal and unchanging. If the one thousand dollar computer didn't exist, one might have thought that people chose computers on the basis of software availability rather than price. The introduction of a cheap computer to the market changes the priorities people assign to different dimensions in making their purchasing decisions. The same is true in other domains as well. Now that the gasoline shortage has passed and prices have stabilized, American consumers don't seem to care very much about fuel efficiency in deciding what kind of car to buy. But let some manufacturer come up with a car that gets one hundred miles to a gallon of gas, and fuel efficiency will become important again.

4. *Economic agents rarely, if ever, act on the basis of complete information.* In order to choose so as to maximize preferences, one must know what is possible. And that means both what is possible now and what will be possible in the future, for if something good is just around the corner (for example, a thousand dollar computer with lots of software), the preference-maximizing choice may be simply to wait. A metaphor for choice with full information is the situation that one confronts when eating at a Chinese restaurant. There, arrayed on the menu, are countless dishes along with their costs. In the closed universe of the Chinese restaurant, complete information is available. One can deliberate about the various possibilities, and when a selection is finally made, it can truly be said to be preference maximizing. But life is not like a restaurant menu; the possibilities for choice are open, unlimited, and ever changing. So perfect information in real life is a myth.

In fact, perfect information is even a myth in the Chinese restaurant. How many people really know what each of the dishes available is like? How often does one study the menu, awed and impressed at the variety available, only to order old favorites? Even in the closed and simple world of the Chinese restaurant, factors other than rational deliberation seem to govern choices. One of them is habit. Even after agonizing over all the possibilities, people fall back, more often than not, on what they have done before. Another factor is tradition. People sit there trying to decide between novel shark's fin and familiar hot and sour soup and finally choose one of them, never considering the possibility that they could have both. One simply doesn't have two soups at a meal. If people fall back on habit and tradition even in a situation where rational deliberation with full information is possible, imagine how much more inclined they are to do so in the situations of everyday life that are full of uncertainty.

Perhaps it is rational to fall back on habit and tradition. Better safe than sorry, after all. It takes time and effort to gather the relevant information, and it is always possible that after it is gathered, old favorites will remain favorites. So one can economize on the costs of information gathering and simply make do with choices that are known to be satisfying, even if they aren't necessarily the most satisfying choices possible.

Modern economics has in fact moved in this kind of direction. Rather than assuming that people possess all the relevant information for making choices, economists treat information as an economic "good," something that has a price (in time or money) and is thus a candidate for consumption along with more traditional goods. Treating information as a good makes the picture of economic rationality more realistic. But a significant problem remains. How much information is it rational to collect before actually making a consumption decision? Consider, again, as a concrete example, trying to decide what kind of personal computer to purchase. How much time should be invested in researching the possibilities? This is a difficult question indeed, for there are so many possibilities, and they keep changing. Computing power keeps going up, while cost keeps coming down. One could make a full-time job out of staying on top of what is available. Complete research obviously is not the solution, at least not if one actually wants to buy and use a computer, but how much less than complete research is rational? There is no way to tell, since how effective the particular strategy chosen turns out to be will depend on how well the chosen model satisfies one's desires *and* on what else is available that wasn't even considered. So treating information as a good, a cost, does not solve the problem of determining what is or is not a rational way to proceed.

6. *Economic agents do not, in general, maximize their preferences.* Owing in no small measure to the information problem just discussed, the economic dictum that people choose to maximize their preferences is almost always false. Indeed, given that information is imperfect, maximization is virtually impossible; when it happens, it is a kind of happy accident of not having failed to consider something important. Even if the environment in which people operate were kind enough to be like a Chinese restaurant menu, that is, with all possibilities and their costs conveniently arrayed to be examined, people simply do not possess the cognitive power to compute the costs and benefits

of all possible combinations of items and so determine which combination maximizes their preferences. For any particular set of possibilities, it may be possible to determine that one is *better* than another, but an examination of the possibilities would need to be exhaustive in order to determine that one possible combination is actually best. And exhaustive evaluation is simply not possible in any choice environment of any complexity. (If this seems implausible then as an exercise, sit down and list all the things one can do with fifty dollars, and then determine which of them will be preference maximizing.) If maximization is impossible, what is the poor economic agent to do? Should he just throw up his hands in dismay and choose at random? Economist Herbert Simon has shown that if we make a few simplifying assumptions about what economic agents are, they can do pretty well indeed, much better than choosing at random. They can, if not maximize, "satisfice." Satisficing is not doing the best one possibly can, but it is doing well enough. What must an economic agent be like in order to satisfice?

First of all, he must have a limited number of goals. The range of things he desires cannot be limitless. Second, for each of these goals, he must have a fixed level of aspiration. He must treat a certain amount of a given commodity as good enough. In short, he must not be endlessly acquisitive. Third, he must have a limited time horizon. He must not be concerned with possibilities that extend into the indefinite future. Finally, he must be able to distinguish clearly means from ends. He must be clear about what his goals are and not confuse the means to achieving those goals with the goals themselves. Indeed, he should be relatively indifferent as to the means of achieving goals.

What each of these characteristics does for a person is restrict the range of possibilities that must be considered before action is taken. Having limited goals obviously achieves this effect. All kinds of things one might choose need not be considered at all if they are not objects of one's desires. Similarly, deciding that a certain amount of something is "good enough" frees one from trying to figure out which strategy will allow him to get the most. All one has to do is consider strategies until he finds one that meets his criterion. Then he can stop considering, unconcerned that other possible strategies might be even better. One also saves deliberation time by having limited time horizons. This saves worry about the consequences of deferring consumption for another five, ten, fifteen, or twenty years. And finally, being clear about the difference between means and ends saves work. All one needs to ask about particular route to a goal is, "Will it get me there?" not "Will I like it more or less than other possible routes?" None of these presumed characteristics of a satisficing person is particularly implausible. The problem is that rational economic agents, as described before, don't possess these characteristics. They have unlimited wants, they want as much as they can possibly get, they project their concerns indefinitely far into the future, and they can have means turn into ends. For the rational economic agent, satisficing does not seem to be an option.

7. *The preferences of rational economic agents are importantly and systematically unstable over time.* Relative stability of preferences is an important component of economic rationality. Were preferences dramatically unstable, we could never distinguish choices that seemed irrational from mere changes in preference, and this failure would make the very concept of economic rationality hollow. Yet evidence is accumulating that preferences are intrinsically unstable. The very act of consuming goods and services changes the satisfaction people derive from them. Much of this evidence has been brought together in a recent psychological theory of motivation developed by Richard Solomon, called the "opponent process theory." Its essence is captured in this remark from Plato, two thousand years ago:

> How singular is the thing called pleasure, and how curiously related to pain, which might be thought to be the opposite of it; for they are never present to a man at the same instant, and yet he who pursues either is generally compelled to take the other; their bodies are two, but they are joined by a single head.

Plato is commenting on how any experience that arouses one emotion in a person seems also to arouse its opposite. In modern dress, in Solomon's theory, this phenomenon is known as *affective contrast*. The idea is this: any experience that causes a deflection—up or down—from emotional neutrality has an aftereffect that is opposite in direction. The body seems designed to keep people on an even keel, and it will work to compensate for any event that steers them off emotional course. What Solomon adds to Plato's doctrine (in addition to confirming empirical evidence) is that the phenomenon of affective contrast is dynamic. A pleasurable experience is followed by a small, unpleasant aftereffect. If the pleasurable experience recurs, its direct, positive effect will remain constant. However, its negative aftereffect will change. It will begin earlier and earlier, even while the positive experience is occurring, it will last longer and longer, and it will grow stronger and stronger. As a result of this change, the negative after-effect will start canceling out the direct, positive effect, so that it doesn't give much pleasure any more. Indeed, the principle emotion a person will feel from this supposedly pleasant experience is the negative one. While having the experience, positive and negative will cancel each other out, so that the person feels roughly neutral. But when the experience ends, he will be left with a strong, negative aftereffect.

To get rid of this aftereffect—to cancel *it* out—people seek out another positive experience. The prime example of this kind of emotional dynamic is drug addiction. When one first takes a drug like heroin, there is a big positive emotional effect and only a small negative aftereffect. But as one continues to take it, the positive effect gets smaller and smaller as it is being canceled by the ever growing negative aftereffect. And when the drug wears off, one is left with a massive negative aftereffect. This aftereffect, usually called *withdrawal,* is what drives people to find another dose. Cigarette smoking is another clear example. If a once heavy smoker who has quit returns to smoking an occasional cigarette, the pleasure he experiences from smoking is enormous. But as he continues to smoke, he derives less and less pleasure from the act of smoking and principally feels the displeasure associated with abstinence.

While Solomon's theory seems to fit the case of drug addiction perfectly, it is not restricted to drugs. The same dynamic process of affective contrast accompanies many experiences of pleasure. After suffering through many a hot and humid summer, one finally scrapes together enough money to buy an air conditioner. What pleasure it brings just to sit still and be cool! But as the days pass, one gets more and more used to being cool. The air conditioning stops giving pleasure. It continues to provide comfort, to ward off the unpleasantness of heat and humidity, but it no longer provides a kick. Then, the air conditioner breaks down. Now, all one feels is intense discomfort—much worse than before there was any air conditioner at all. Even sleeping at night is difficult.

What this example describes is a familiar phenomenon. It describes how luxuries become necessities. One of the things that makes something a luxury is the substantial pleasure that it provides. But as people get used to it, it stops giving pleasure, and what they principally feel is the displeasure of having to do without it. Thus, air conditioners, cars, telephones, televisions, washing machines, and the like seem to many to be absolutely essential for existence. Soon, dishwashers and video recorders will join the list of necessities. And not too far down the road, word processing computers will be added to the list.

What does all this have to do with changing preferences? Economist Tibor Scitovsky has argued in his book *The Joyless Economy* that mere comfort is not good enough for people; they also want pleasure. However, they cannot keep deriving pleasure from consuming the same old things. There is an inevitable disappointment that comes with consumption because repeated consumption of the same commodities provides people with doses of pleasure that do not live up to their expectations—expectations shaped by their initial encounters with those commodities. As a result, people are driven to pursue novelty, to seek out new commodities whose pleasure potential has not been driven down by repeated consumption. But these new commodities will also eventually fail to satisfy. The lesson in this is that the pursuit of pleasure is a perpetual wild goose chase. It requires people to be always on the lookout for new things. This may help explain the seemingly irrational and self-destructive thrill seeking that seems to characterize some especially affluent members of our affluent society.

In support of his argument for the inevitable disappointment of consumption, Scitovsky cites the results of surveys of how happy Americans thought they were over a twenty-five-year period from 1946 to 1970. During this period, real (inflation-adjusted) income in the United States rose 62 percent. So people on the whole were much better off in 1970 than they were in 1946. Yet this large change in material welfare had absolutely no effect on happiness ratings. People were no happier in 1970 than they were in 1946, although if you had asked them, in 1946, how happy they would be if they had the standard of living that they actually did have in 1970, nearly everyone would have been ecstatic.

This last fact can be understood in two different ways, but both of them pose problems for economics. It could be that happiness does not increase with increases in material well-being just because of affective contrast. Material well-being brings comfort, but comfort is not pleasure and, thus, is not happiness. The problem that this interpretation poses for economics is that it means that preferences will always be changing. On the other hand, it could be that material well-being is relevant to happiness only when it is evaluated *relative* to the material well-being of everybody else. A 50 percent increase in real income will make someone happy only if not everyone's income has increased by 50 percent. If everyone is getting richer, then an individual's own gains are seen as only fair—as entitlements. They don't change his relative position in society at all, and thus they don't make him happy. It is certainly not implausible that for many people what matters is not simply how much they have but how much *more* they have than others. And this could account for why societywide changes in affluence do not affect individual judgments about happiness.

This interpretation would save economics from the problem of ever-changing preferences. However, it would substitute another, no less important problem. We saw in Chapter 3 how economists attempt to evaluate how well economic developments serve the interests of society as a whole. Owing to interpersonal incommensurabilities of preference, it is very difficult to make this assessment in situations in which one person's gain of X means another person's loss of Y. But at least economists can identify as improvements in welfare economic changes that satisfy the Pareto criterion. If at least one person is made better off without anyone being made worse off, welfare has been increased.

However, if, as suggested here, a person's own assessment of his welfare is made relative to what other people have, then no economic change can ever satisfy the Pareto criterion. Any change that makes one person better off will make others worse off, not because the concrete conditions of their existence have changed but because their relative position has changed—for the worse. Thus, imagine someone earning forty thousand dollars a year, with a nice house, two televisions, a station wagon, and enough loose change in his pocket for an occasional indulgence. If everyone else around him is making about the same amount and living in about the same way, he may well be quite satisfied. But now suppose that the town discovers oil and the town elders decide to give everyone but him a million dollars. This decision certainly satisfies the Pareto criterion; several people are being made better off, and no one is being made worse off. Right? Wrong! As he sees people putting swimming pools in their backyards and trading their station wagons for Mercedes, his level of satisfaction with his own lot starts to sink. Before long, he starts feeling deprived, even poor. So the Pareto criterion hasn't been satisfied after all. The distribution decision has indeed left one person worse off. If, in general, people evaluate their own welfare in relative terms, relative to what other people have, then any improvement in one person's welfare will diminish everyone else's. In effect, no change in the distribution of resources will ever satisfy the Pareto criterion, save a change in which all benefit equally. People have always said that money can't buy happiness. Among those who said it was none other than Adam Smith:

> Power and riches appear then to be . . . enormous and operose machines contrived to produce a few trifling conveniences to the body. . . . Though they may save [one] from some smaller inconveniences, [they] can protect him from none of the severer inclemencies of the season. They keep off the summer shower, not the winter storm, but leave him always as much, and sometimes more, exposed than before to anxiety, to fear, and to sorrow; to diseases, to danger, and to death.

People have always said it. Now we know why they are right.

ECONOMIC IRRATIONALITY: A SUMMARY

We have now spent a good deal of space identifying the ways in which economics presents an inaccurate, or at least limited picture of the way people actually are and how they actually operate. Real people, unlike economic men, cannot express preferences among all possible commodities, do not always prefer cheap to expensive, do not always have transitive preferences, do not act with complete information, do not, indeed cannot, act to maximize preference, and do not have stable preferences. Furthermore, the extent to which people do approximate economic rationality is heavily influenced by factors that are themselves noneconomic in nature. One's ability to seek relevant information, to manage the information that he gets, and to restrict the infinite number of possible choices he can make to a finite and coherent set depend upon a network of traditional cultural practices and individual habits about which the economic framework is silent. The idea of economic rationality is on the one hand too rich, by giving people credit for more calculation and consistency than they possess, and on the other hand too impoverished, by failing to appreciate a range of noneconomic influences on economic decision making. It is hard to avoid the conclusion that economic rationality is not so much an abstraction—an idealization—as it is a creation—a fiction.

The Framing of Decisions

Will people buy gasoline at a gas station that offers a discount for paying in cash? How about a gas station that imposes a surcharge for using credit cards? Does the first of these options seem like a good deal while the second seems like a ripoff? That's the way many people see it, although in fact, the two options are identical. A discount for paying in cash *is* a surcharge for using credit. Rational economic agents would know this. They would not be affected by whether options were presented as opportunities to save money or as opportunities with added costs. They would respond only to the net cost—the bottom line. But an ingenious series of investigations by Daniel Kahneman and Amos Tversky over the last decade has shown that most people are consistently and systematically irrational in situations like these. They fail to see past the superficial way in which alternatives are framed to their underlying equivalence. What people choose is affected by the way in which alternatives are presented to them.

Consider being posed the following problem:

> Imagine that you have decided to see a play where admission is ten dollars a ticket. As you enter the theater you discover that you have lost a ten dollar bill. Would you still pay ten dollars for a ticket to the play?

Almost 90 percent of people asked this question say yes. In contrast:

> Imagine that you have decided to see a play and paid the admission price of ten dollars a ticket. As you enter the theater you discover that you have lost the ticket. The seat was not marked and the ticket cannot be recovered. Would you pay ten dollars for another ticket?

Now, less than fifty percent of people say yes. What is going on here? What is the difference between the two cases? From one perspective, they seem the same; both involve seeing a play and being twenty dollars poorer or not seeing it and being ten dollars poorer. Yet people don't seem to see them as the same. What Kahneman and Tversky suggest is that the difference between the two cases has to do with the way in which people keep psychological accounts. Suppose that in a person's internal accounting system there is a "cost of the theater" account. In the first case, the cost of the theater is ten dollars; the lost ten dollar bill is not properly charged to that account. But in the second case, the cost of the theater is twenty dollars (two tickets), and for many people, twenty dollars is too much to pay. On the other hand, suppose that the person's internal accounting system has a "cost of a day's outing" account. Now, the two cases may well be equivalent, in that the lost ticket and the lost ten dollars both add the same amount to the cost of the day. So some people keep narrow, cost of the theater accounts, while others keep broader, cost of the day accounts. Which of them is rational? What is the way in which rational, economic agents *should* keep their accounts?

The economist's answer to this question might be that the rational way to keep accounts is whatever way allows one to maximize preferences. If it makes someone feel better to keep the lost ten dollars in a separate account from the theater costs, then she should do it. If not, then she shouldn't. If it makes someone feel better today, she should do it today. Tomorrow she can keep accounts differently. The problem with this answer is that since there is no independent way to know whether or not people are maximizing their preferences, we are left assuming that whatever accounting system they are using is just the one that serves their preferences best. Furthermore, given the enormous diversity and fluidity of possible accounting systems, it can never be asserted with any confidence that a given choice is or is not rational. For whether it is rational will depend on how accounts are being kept. In short, in the absence of a specification of either what accounting systems people should use or what accounting systems they actually do use, the notion that people act as rational economic agents is of little help in predicting what they will do.

And the range of possible accounting systems really is enormous. For example, a journey to the theater could be just one entry in a much larger account, say a "meeting a potential spouse" account. Someone might go to the theater, the ballet, and so on because he hopes to meet his type of person there. Or it could be part of a "getting culture" account, in which case it would be one entry among others that include subscribing to public television, buying certain books and magazines, and the like. It could be part of a "ways to spend a Friday night" account, in which case it would join entries like drinking at a bar, going to a basketball game, figuring one's income tax, and who knows what else. How much this night at the theater is "worth" will depend on what account it is a part of. Twenty dollars may be a lot to spend for getting culture, compared with available alternatives, but not much to spend to find a spouse or pass a Friday night. In sum, just how well this twenty dollar night at the theater will satisfy one's preferences is going to depend on how he does his accounting.

And this is a very important fact. What it means is that events or commodities don't have absolute values in and of themselves. We can't say much about the value something has just by knowing how much someone was willing to pay for it. The same ten dollar pay-out may reflect trivial value in one accounting system and significant value in another. The value of things depends on the context in which they are evaluated.

We often talk jokingly about how "creative" accountants can make a corporate balance sheet look as good or as bad as they want it to look. Well, the point here is that people are all creative accountants when it comes to keeping their own psychological accounts. This very creativity raises an important question. If there are no norms or standards of rationality to judge accounting systems by, and if the number of possible accounting systems really is indefinitely large, what is it that determines which accounting systems people actually use? We can't say they use the rational ones, since we can't say which ones are rational. What, then, can we say?

In approaching this question, a look at "real" accounting practices can be instructive. Real accountants can also organize accounts in indefinitely many ways. What constrains the way they operate? There are three sources of constraints. One source is legal. Tax laws impose a set of requirements on how the books must be kept. A second source is professional. The accounting profession establishes certain standards that guide how accounting is to be done. They maintain those standards in part by educating new accountants to do things in just that way. The final source is customary or habitual. Accountants keep accounts in certain ways because they have always kept them in those ways, or because the accountants who preceded them kept them in those ways. There is nothing especially privileged or rational about these constraints. Tax laws could be different, as could professional standards, and habits are accidents of history. So in a sense, these constraints are arbitrary, that is, unjustified. Yet, arbitrary or not, they are there, and they serve to narrow and shape the way accountants do their work.

Precisely the same things could be said about the way people keep their psychological accounts. They are influenced by legal and social sanctions, by customs and traditions, and by old habits. These influences may also be arbitrary—unprivileged and unjustified. Nevertheless, people inherit them and their effect on the keeping of accounts. People don't include their income tax or the cost of supporting their children in their "charitable giving" account, though they could. They don't keep food costs in a "medical expenses" account, though they could. They don't keep college tuition and movie theater admissions in the same "educational costs" account. They don't treat school taxes as child care expenses. They don't treat the money they give to houses of worship as entertainment costs. People have good reasons for not doing these things, but they are not economic reasons. They stem from being influenced by their culture as to what categories to establish and what items to put in each category. Psychological accounting practices in different cultures will be quite different from ours, but just as reasonable.

There is an extremely important implication for our evaluation of economic theory embedded in this fact about psychological accounts. We have seen that while economists concede that "economic man" is not total man, it is autonomous of the other, noneconomic aspects of human nature. It is clear now that this claim is false. If an assessment of what is rational depends upon the accounts that people keep, and the accounts that people keep in turn depend upon noneconomic habitual or cultural practices, then an assessment of rationality depends on those practices. Rather than being autonomous, the economic portion of human nature is through and through dependent on the culture in which economic practices are embedded.

And in some ways, these cultural constraints on economic activity are a good thing economically. They may be what makes it possible for economic activity even to approximate rationality. We have seen that perfect rationality is impossible because people are unable to state preferences among all different possible alternatives or to gain complete information about all present and future prospects. Given these limits on the ability to get information and make comparisons, how do people decide what information to select and what comparisons to make? If these decisions were simply made at random, economic activity would be chaotic and unpredictable—anything but rational. What culture and habit do is ensure that people don't select at random. They select in keeping with cultural norms, in more or less the same way that others select. This is what makes one person's economic actvity intelligible to others, and the economic activity of others intelligible to him. This is what makes it possible for actors in the economic marketplace to "read" the market and make intelligent decisions about what people will be willing to buy and how much they will be willing to pay for it. Cultural constraint on accounting is one limit on economic rationality that makes all other approximations to economic rationality possible.

What Is Ecomomic Activity?

To the economist, economics is about the allocation of scarce resources, and economic activity is the exchange of those resources in the marketplace. This idea seems reasonable on its face. It probably comes pretty close to matching most everyday conceptions of what counts as economic activity. Economics is about buying and selling. It's about managing financial resources and trying to come out ahead in most transactions. What else could it be?

To see what else it could be, imagine a small farmer, prior to the industrial revolution, say, three hundred years ago. What would his economic activity have been like at this time? For the most part, it wouldn't have been exchange in the market, for there were almost no markets, and what markets there were rarely reached very far afield, given the limits on available transport at the time. Does this mean that the preindustrial farmer engaged in little economic activity? And by extension, does this mean that people living presently in preindustrial societies engage in little economic activity? If so, then economic rationality, even in limited form, can hardly be a characteristic of human nature. If ecomomic activity is exchange, and exchange depends upon markets, and the extensiveness of markets depends upon industrialization, then economic rationality is the product of the particular series of historical and cultural developments that established markets.

On the other hand, perhaps economic activity can be viewed as something other than exchange. Perhaps it can be described more generally, so that it encompasses both the free marketeer and the small, preindustrial farmer. We can do this by returning to the meaning that *oeconomics* had to ancient Greeks. Economic activity is the activity that sustains life. Economic activity is production. It is the production of daily life. For the farmer, this means raising crops, keeping chickens for eggs and cows for milk, doing occasional hunting and fishing, skinning animals for clothes, spinning cotton or wool, keeping the farm buildings and machinery in repair, caring for the plow horses, and so on. Not an item of exchange in the lot. For the modern white collar worker, this means exchanging time administering the activity of underlings, writing memos, and so on, for money, using that money to make wise investments, exchanging it for an accountant's advice that will help in avoiding taxation, and exchanging it on the market for desired commodities. Not an item of production in the lot. Only in the second case does the production of daily life involve exchange.

It might be tempting to argue that the distinction between production and exchange is not a real one. The preindustrial farmer *is* engaged in exchange. He is exchanging his labor time for goods instead of money, but it is a process of exchange nonetheless, no different in principle from the activity of the white collar worker. But if we try to take this argument seriously and apply rational economic man concepts to the activity of the farmer, most of them don't make much sense. The amount of time that he spends at his various tasks cannot be treated as a measure of the value of their products to him. Farming may take ten times as much effort as hunting. From this, it does not follow that vegetables are ten times as valuable as meat. The farmer needs them both, and the time he spends at his various activities is dictated by the demands of the activities themselves and not by any calculation of value. Indeed, it is hard to know how to begin to assess value in economic terms. Value for economists is exchange value. It is how much of one thing one is willing to give up for how much of another. Nothing of this sort can be determined in the case of the farmer. Nor is it clear when one is living a life that is entrained to the rhythms of nature that one is ever in a position to be making choices that maximize preferences. The categories of economic rationality are just the wrong categories for understanding what the farmer does. Certainly, there can be better and worse farmers, rational and irrational ones, but rational farmers and rational economic men are not just two sides of the same coin.

Thinking about economic activity as production rather than as exchange leads one to think more seriously about economic activity *as* activity. It leads one to ask how people spend their days, what the nature and the point of human work is. With economic activity regarded as exchange, the "activity" gets lost. No one cares whether trading on the market is done by mail order or by shopping in person. No one cares about the character of the work a person does, since work is just the means to consumption. The presumption is that people sell their labor to the highest bidder, to maximize the rate of return on labor time, and thus maximize opportunities to consume. An employer will dictate what the worker does so as to maximize his rate of return, or profit. So efficiency seems to be the critical dimension in determining the nature of work. Both worker and employer want to get the most return out of the least effort.

When work is regarded in this way, simply as a means to consumption, it is a pure disutility. The ideal for someone would be to be of independent means so that he could consume all the time without having to sell his labor. Given the nature of the jobs that many people have in modern societies, thinking of work as a pure disutility may not seem unreasonable. Few people would do what they do without substantial compensation, and most of them would be happy to sell their time to the highest bidder. But things needn't be this way. Some people enjoy what they do and wouldn't trade jobs simply for higher wages. Is it possible to organize work so that almost everyone achieves some satisfaction from his daily activities?

At first glance, it seems that there are things that must be done in society that no one likes to do. Not everyone can have a meaningful, satisfying job. While this may seem true, on closer inspection the difference between meaningful and menial is not always obvious. Mindless, backbreaking work like moving boulders and clearing a field may seem a pure disutility, and no doubt, if that were all one did, it would be. But as part of a larger project, like planting a garden or building a house, it becomes meaningful. Pushing heavy objects around and running around in circles seem meaningless if anything is. But weight lifting and running to get in shape for the football season are not meaningless. So if work were organized so that menial elements were part of a larger, meaningful project, perhaps everyone would get satisfaction out of work.

Why, then, isn't work organized in this way? One economic answer is that it would be inefficient to organize work in this way. Productivity is increased if work is broken down into highly specialized components. But this answer has built into it the idea that consumption is the point of life and work is a disutility. Nonspecialized work is inefficient if the point is to produce as many commodities as possible for as low a price as possible. But if we assume instead that the point, at least in part, is to gain some satisfaction out of work, consider the inefficiency, the cost, of having people spend a third of every day engaged in menial labor. This is a cost that is difficult to measure. It cannot be easily translated into dollars. So for most economists it is conveniently ignored. Indeed, both the cost of menial work and the benefit of meaningful work are ignored. Neither is a commodity. Both are external to economic activity.

The economist who argued most forcefully that thinking about work as a pure disutility was exactly the wrong way to look at things was Karl Marx. For Marx, work was the essential ingredient of being human. By working, people transformed nature. They made the world over in their own image. This was what gave human life meaning. Work was the point of life. People ate in order to work. To the economist, people work in order to eat, to consume. This view was termed by Marx *commodity fetishism.* He saw it as the inevitable product of industrial capitalism, as the strongest sign of how the market system could pervert human potential. This idea is well captured by the words of philosopher Albert Camus: "Without work all life goes rotten. But when work is soulless, life stifles and dies."

But wait a minute. If the quality of work really was important to people, they would not sell their time to the highest bidder. They would be willing to forgo some wages in return for satisfactory working conditions. Why don't they? Why are labor–management contract negotiations so often about wages and benefits and not about working conditions? There are several answers to this question. First, workers have little control over what the opportunities are. If all available jobs are more or less the same in character, why not sell one's labor to the highest bidder? Second, many issues related to working conditions have traditionally not been allowed on the table in labor–management negotiations; there have been severe limits to what management was willing to negotiate about. What the worker actually does for his wage has been viewed as entirely subject to management's discretion. And finally, living in a society in which work is assumed to be a disutility and the workplace is designed in a way that is consistent with that assumption can affect one's aspirations—one's conception of what the possibilities are. To see this, recall Adam Smith's cautionary note about the consequences of extreme division of labor:

> The man whose life is spent in performing a few simple operations . . . has no opportunity to exert his understanding, or to exercise his invention. He naturally loses, therefore, the habit of such exertion and generally becomes as stupid and ignorant as it is possible for a human creature to become.

Rational Economic Man and History

How does it come to be that one's principle economic activity, the production of one's daily life, is dominated by exchange? What turns the preindustrial farmer into a white collar or assembly-line worker? Exchange starts to become essential when people stop producing for themselves everything that they need in order to live. When the subsistence farmer becomes a cash farmer, he must enter the market-place. Man cannot live by bread (or eggs, or milk) alone. This is all the more true when one is involved in manufacturing. The need for exchange is obvious when one looks at modern, urban life. Virtually no one who lives and works in a city produces what he needs to stay alive. If it weren't for exchange in the market, city dwellers would all starve to death. But there are all kinds of possible gradations in the extent to which exchange dominates economic activity that lie between the preindustrial farmer and the modern urban dweller. Imagine, for example, the craftsman who makes leather goods but also has a small garden and a few animals. Much of his food is gained not through exchange but through production. However, he does exchange his leather goods for many items that he does not produce himself. Perhaps in the modern world, the decision about what to produce for oneself and what to acquire by exchange is a reflection of economic rationality, but two hundred years ago, it was probably the result of deeply ingrained customs that governed how people lived their lives, together with constraints imposed by what was actually available through exchange.

What has largely eliminated many of the gradations between pure production and pure exchange is the industrial revolution that began in the seventeenth century. It brought with it an increase in special-ization of work, so that very few people engaged in production actually produced anything of use in itself. Each worker produced only a part of some useful thing. It also took people away from the home and into the factory, making it difficult to engage in subsistence farming and production for exchange at the same time. So the notion that eco-nomic activity is exchange, and the development of markets in which practically anything can be exchanged, are very much products of the industrial revolution. This makes rational economic man, as described by economists, an entity that exists under only a rather restricted set of conditions that have obtained only in the recent history of our species, and then only in certain parts of the world.

This argument has been eloquently and persuasively made by eco-nomic historian and anthropologist Karl Polanyi. In his book *The Great Transformation*, Polanyi traced the development of the market system in Western Europe, contrasting the market as an economic and social institution to the social institutions it replaced. For people like us, who have been born and raised into a particular system of social insti-tutions, it is often hard to imagine a society running in any other way. It is hard to conceive of an alternative to the market. That is why a look at history, and at other cultures, can often be so enlightening. Polanyi's argument about the development of the market system can best be summarized in his own words:

> No less a thinker than Adam Smith suggested that the division of labor in society was dependent on the existence of markets, or, as he put it, upon man's "propensity to barter, truck and exchange one thing for another." This phrase was later to yield the concept of the Economic Man. In retrospect it can be said that no misreading of the past ever proved more prophetic of the future. For while up to Adam Smith's time that propensity had hardly shown up on a considerable scale in any observed community, and had remained, at best, a subordinate feature of economic life, a hundred years later an industrial system was in full swing over the major part of the planet which, practically and theoretically, implied that the human race was swayed in all its economic activities, if not also in its political, intellectual, and spiritual pursuits, by that one particular propensity.

To support this claim, Polanyi discusses evidence gathered by social anthropologists in the study of various preindustrial cultures. The principal concerns of people in these cultures were not the pursuit and safeguarding of economic self-interest but the pursuit and safeguarding of social standing, of membership in the community. Thus, in studies of the "economics" of societies such as the Trobriand Islanders of Western Melanesia, one finds that reciprocity and redistribution, and not selfish accumulation, governed the activities of individuals. The fruits of individual labor were placed in common storehouses, to be redistributed as needed. Decisions about redistribution were not in general democratic, and they may not have always been especially fair. They were typically in the hands of the tribal chief or the priest. However, distribution decisions were always quite clearly political and social decisions, and they were *decisions*. Distribution was not left to natural forces, as it is in the case of the free market. That reciprocity and redistribution governed economic activity does not mean that the individual worked tirelessly and effectively out of entirely unselfish motives. A great deal of social standing was at stake for a man and his family based upon his skill as a farmer. So it was in his *social* interest to be a good one. But one's social standing in the community bore little or no relation to one's economic standing.

This lack of interest in accumulation is a persistent theme in studies of preindustrial cultures. Thus Marshall Sahlins points out, in his book *Stone Age Economics,* that in primitive agricultural societies, people farmed far less of the available land than they could. They produced only what they needed, and no more. And similarly, as Polanyi and others have pointed out, European colonialists were consistently frustrated in their efforts to get natives to work hard in the factories that were established in various colonial locations. Efforts to spark incentive by increasing wages typically led to a decrease, rather than an increase, in the natives' willingness to work. The reason for this is obvious, though it may have been incomprehensible to the European, who assumed that people were driven to make as much money as they could. Natives worked until they had what they needed; then they stopped. As wages were increased, it took less work to satisfy their needs. The way to get natives to work hard, it was eventually discovered, was to keep them on the brink of starvation.

The members of such preindustrial societies even engaged in trade, so Adam Smith's claim about man's propensity to truck and barter is not entirely off the mark. But trade had an entirely different character than it does in the market. It was governed, not by the pursuit of gain, but by reciprocity, and very personal reciprocity at that. Each individual in one community might have his own personal trading partner in the other. So while the goods of the communities might move back and forth en masse, this massed exchange was simply a convenient means for accomplishing many, many personal exchanges. This is rather a far cry from the anonymous marketplace of modern ecomomic systems.

In addition to the principles of reciprocity and redistribution that governed trade and production for the common pool, Polanyi identified one other characteristic of premarket economic activity, which he termed *householding.* Householding is production for use, the case of the preindustrial, subsistence farmer already described. Interestingly, Polanyi traces the idea of householding, and the distinction between production for use and production for gain, to Aristotle. Aristotle argued for production for use and against production for gain. Production for gain was all right if it was marginal, if a family simply sold its surplus on the market. But if one started producing exclusively for gain, if one started raising crops for sale that one otherwise wouldn't raise at all, Aristotle thought that the self-sufficiency of the household would gradually but inexorably erode. Aristotle, in contrast to Adam Smith, thought that production for gain was not natural to man and that, once begun, it would be boundless and limitless. The end result of production for gain would be the divorce of economic motives from the social structure and social relations in which they were otherwise embedded. Thus, what economists describe as the "autonomous economic motive" was, for Aristotle, the result of production for gain, and not a precondition for it.

According to Polanyi, these characteristics of economic activity that are now observed in preindustrial cultures were also there to be observed in preindustrial Europe, at least until the seeds of the market system were sown in early seventeenth-century England. At this time, the great majority of people were peasants who eked out subsistence by farming common lands. They were tied to these lands and to the communities of which they were a part. This mode of existence was changed by enclosure laws, which did away with common lands, giving title to these lands to the English nobility. This enclosure of farmland had two effects. First, agricultural production became much more efficient, with the output per acre often doubling or tripling. Second, thousands of peasants were driven off the land and left to wander the countryside, with no means of support. The dislocated peasant farmers were to provide the labor force that started filling factories a century later.

At roughly the same time, markets started coming into existence, made important by the growing capabilities of society to engage in long-distance trade. Long-distance trade required markets for local distribution. The existence of markets, in turn, engendered exchange, buying and selling. But these early markets were nothing like the free, self-regulating ones of modern economic theory. A self-regulating market is governed only by price. Production and distribution are regulated by the forces of supply and demand operating on people out to maximize their gains. Order, efficiency, and productivity result, and no forces exogenous to the price system are necessary. The factors operative in a market economic system are goods, land, labor, and money. If the market system is fully functioning, each of these factors must have its price, in commodity prices, rents, wages, and interest. And each of these prices must be left completely free to vary, determined only by the interplay of supply and demand in the market. Neither the state nor any other noneconomic social institution must be allowed to interfere.

In fact, when the market system began, virtually none of these conditions obtained. Land, for example, was often inalienable, that is, unsellable, and if transfers of land were permissible, who could engage in the transfers, and under what conditions, was rigidly constrained by social rules. The uses to which land could be put were also heavily regulated. The same was true of labor. Under the guild system, one's occupation was largely inalienable; a man did what his father had done, in the way that his father had done it. And he did not auction his services to the highest bidder; instead, he moved through a highly structured system, from apprentice to journeyman to master, with wages at every step of the way determined not by the market but by social custom. What a man did, how he did it, and what he got for it were determined by traditional rather than by market forces. Thus, neither labor nor land were truly objects of commerce. And interest, then known as usury, was commonly prohibited all together. So, at least at its inception, the free, self-regulating market was hardly either free or self-regulating; it was constrained right and left in its operation by other social institutions. Eventually, and perhaps inexorably, as Aristotle foresaw, these various constraints eroded, and increasingly close approximations to what economists now regard as the free, self-regulating market emerged. Polanyi puts it this way:

> The peculiarity of the civilization the collapse of which we have witnessed was precisely that it rested on economic foundations. Other societies and other civilizations, too, were limited by the material conditions of their existence—this is a common trait of all human life, indeed, of all life, whether religious or non-religious, materialist or spiritualist. All types of societies are limited by economic factors. Nineteenth century civilization alone was economic in a different and distinctive sense, for it chose to base itself on a motive only rarely acknowledged as valid in the history of human societies, and certainly never before raised to the level of a justification of action and behavior in everyday life, namely, gain. The self-regulating market system was uniquely derived from this principle.

Economic History and Economic Science

The critical message in Polanyi's analysis is that economic men are made, not born. The propensity to truck and barter is not "natural," not a part of human nature. Neither is the desire to exchange for gain. However, both of these characteristics can be a part of human nature under the right conditions. To see economic man in full flower, one needs a society organized around an unfettered market, in which social institutions that might restrict exchange of land, labor, money, and goods have been systematically eliminated. The market system is not made possible by economic men; rather, it makes economic men possible. The implications of this line of argument for economics are significant. Remember, in the eyes of economists, economic concepts are not mere descriptions of particular points in history. They are scientific laws, fundamental truths about the human organism and the human condition.

What, exactly, is a scientific law? As Charles Dyke points out in his book, *Philosophy of Economics,* one way of thinking about laws in general is as constraints on human activities. The law of gravitation is one such constraint; it keeps people from flying about uncontrollably. The law that prohibits going through red lights is another such constraint; it keeps people from driving their cars in whatever way they like. But these two kinds of "laws" are obviously very different. The constraint imposed by gravity is not manmade, not self-imposed, and it cannot be repealed no matter how much people want to. The constraint on going through red lights, in contrast, is self-imposed and easily repealed.

Which of these kinds of constraints are described by the "laws" of economics? According to Dyke, and to Polanyi, they are clearly like traffic laws, not gravity. Even the "iron law of supply and demand" is the product of human discretion. It presupposes trading, which presupposes the division of labor. It presupposes a market, and a market with particular characteristics and particular rules about what can be bought or sold, and by whom. All of these features of society that make the law of supply and demand ring true are human creations—creations that could be different, that indeed once were different. One could imagine society moving in the future in directions that undercut some of these conditions on which the law of supply and demand depends. One could also imagine society moving in directions that permit the extension of supply and demand to aspects of human life, like marriage and child raising, that they presently do not touch. But either of these moves, should they develop, will be the product of human discretion, not of natural necessity.

If economic laws are like traffic laws, they are immediately in need of justification or defense. Gravity requires no defense; it simply is. Not so for traffic laws. We must defend the infringement on individual freedom they represent. It must be argued that *this* set of traffic laws, and not some others, is the right one. Now such a defense may not be difficult to make, but it will appeal to such things as values and morals; it will depend upon some understanding of what is good for people and what is good for society. The free market can be defended in the same way. People can attempt to justify it by appealing to human rights, freedoms, and entitlements. People can attempt to justify it by appealing to the goods, both social and individual, that will derive from it. That is to say, the free market and economic rationality can be defended—or attacked—on moral grounds. But it cannot be defended as just another gravitational constraint on human activity. And this state of affairs must ultimately be unsatisfying to the economist, whose aspiration was to create an economic science that removed issues like these from the domain of moral discourse.

There is one move open to the economist who wants to attempt to preserve the status of his principles as eternal scientific laws. He might want to suggest that people may not have appeared to be economic men, motivated by self-interest, in preindustrial cultures. Deep down, however, they were, just as much so as in modern, industrial society. Perhaps it is hard to penetrate to the essential, economic nature of preindustrial men because we can't decipher what their wants and preference structures were. Or perhaps their economic natures were suppressed by authoritarian state and religious institutions. Or perhaps it took the evolution of a market system—a progressive step in the history of human culture—to release their essential, economic nature. The economist needs one of these stories, or something similar, to be true, if he is to keep the nature of human nature squarely in the domain of scientific discovery. For aid and comfort in this effort, he can turn to modern evolutionary biology. Thus the concepts of evolutionary biology we have already reviewed may have a significant role to play in defending the scientific status of economics. Let us turn, then, to a critical examination of the biologist's conception of human nature.

The Limits of Evolutionary Biology

*Men unmodified by the customs of particular places do not in
fact exist, have never existed, and most important, could not,
in the very nature of the case exist.*

CLIFFORD GEERTZ

*Before there can be natural selection, there is cultural selec-
tion: of the relevant natural facts.*

MARSHALL SAHLINS

*If the misery of our poor be determined not by the laws of
nature, but by our institutions, great is our sin.*

CHARLES DARWIN

*M*odern evolutionary biology invites us to
understand many of the details of human, social life as the product
of biological determination. It invites us to see human sexuality, parenting, aggression, and altruism as continuous with the determinants
of similar behavior patterns in nonhumans, driven by the logic of
individual fitness maximization. To accept the invitation is to embrace
"economic human nature," broadly construed, not as an historical
fact that could be otherwise, but as a natural law. To accept the
invitation is to move in the direction of a natural morality, to start
treating many aspects of human behavior that are usually regarded as
matters of good and bad instead as matters of fact. This is, in short,

a very significant invitation. Should we accept it?

This chapter will examine the evolutionary biological framework critically and suggest that, although the argument is seductively clever, it should not be accepted. It should not be accepted for two different sorts of reasons. First, there is reason to question the adequacy of the framework even in connection with the phenomena of animal behavior that it was directly designed to capture. And second, with respect to human behavior, there is reason to believe that the biological framework is importantly incomplete, that it ignores or underplays influences on human action that are absolutely crucial.

Evolutionary Biology and Animal Behavior

How good a job does evolutionary biology do of accounting for animal behavior? On the face of it, this would seem like a fairly easy question to answer. Sociobiology claims to be a modern, up-to-date branch of evolutionary biology. Evolutionary biology is science. Science is nothing if not a set of definite propositions about how the world is. Either objects fall when dropped or they don't. Either water is made up of hydrogen and oxygen or it isn't. Either organisms act to maximize inclusive reproductive fitness or they don't. It seems that to evaluate the sociobiological picture all one needs to do is to go out into nature and start collecting facts. The facts are out there, ready to be collected. Either the theory fits the facts or it doesn't. Making the determination may take some legwork, but it should be a reasonably cut-and-dried affair.

If only it were that simple. Most people, including scientists, seem to hold the belief that facts sit out there in the world, like isolated droplets of nectar, waiting to be collected. Our senses tell us, clearly and unambiguously, what the facts are. There may be disputes about how best to account for the facts, about which of a set of theories is most accurate, most plausible, most comprehensive, but there aren't any disputes about the facts. This view is simple, but it is false. As philosophers of science have been saying for decades, there is no clear line between facts on the one hand and theories on the other. Theories not only provide an account of the facts, they indicate what is to count as a fact as well. Rather than isolated droplets, the senses yield rivers of nectar, rivers that, as Heraclitus said, are never the same at two different points in time. People impose some structure on those rivers; they break them up into droplets. But there are countless different ways they can do the breaking. And how people decide to structure the flowing river will determine what they take to be facts.

Partly, what influences the way people structure the world is the theories about it that they hold. One theory of the universe says that it is unending and that the patches of light in the sky at night are stars. Another theory says the universe is closed and that the patches of light are cracks in its outer shell. There is no way to choose between these theories on the basis of facts given by the senses, for what the facts are depends on what theory one happens to hold. Another way of saying essentially the same thing is to note that theories are always underdetermined by the available facts; that is, many different theories can account for a given set of facts. Much of the criticism of sociobiology as an account of animal behavior follows these lines. It is not that sociobiology is patently false—inconsistent with facts. It is just that there are other, equally compelling stories that could be told.

If this bedrock uncertainty about scientific theories exists, how does one ever intelligently choose one theory over another? Is it simply a matter of personal taste, or are there criteria to which one can appeal? There are criteria, and one of them is how well a particular theory connects up with other accepted theories. It is this connecting up with other theories that makes sociobiology such a strong candidate for allegiance. It is of a piece with economic formulations that many people take to be self-evident. It offers the opportunity to understand human behavior in the marketplace, chimpanzee behavior in the jungle, and bee behavior in the hive, with a single, comprehensive explanatory framework. How neat and satisfying such a comprehensive framework would be, if only sociobiology could pull it off.

Critics of sociobiology contend that it cannot pull it off, that in striving for comprehensiveness it makes assumptions about the nature of evolution that are radically incompatible with a proper understanding of evolutionary theory. Let us look at the central claim of sociobiology with some care. *Organisms act so as to maximize utility, with utility defined as inclusive reproductive fitness.* Now this is an extremely powerful claim, full of predictive implications. One could actually compute mathematically what maximization of utility in a situation would look like and then ask whether organisms act the way the mathematics say they should. One could wear an engineer's hat and construct model, utility-maximizing creatures, then ask whether real creatures look and act like the models. What doing either of these things would reveal is that organisms virtually never act to maximize utility.

What can this mean? Isn't a potential mother maximizing utility when it finds the fittest male as mate? No, it could do even better by producing as many eggs as males produce sperm, or by being able to care for its young on its own, or by giving birth to young that require no care to begin with. Isn't the male elephant seal that dominates the entire social group and impregnates all the females maximizing utility? It seems he is, but then why aren't all animal social groups organized into harems the way seals are? And isn't the female salmon maximizing utility when on her one trip upstream to lay eggs, before she dies, she contorts her internal organs to produce as many eggs as possible? Yes, given that she is going to die. But a cleverer female would see to it that this was just one trip upstream among many.

So organisms typically don't maximize reproductive fitness. It is always possible to design a hypothetical creature that could do better. The point is very obvious, and sociobiologists know it and acknowledge it. What is meant by maximization, they say, is not the best possible strategy but the best strategy given certain constraints. These constraints are imposed in part by the animal's evolutionary history. It might be wonderfully effective for a cockroach to produce an offspring that is as big as an elephant, as fast as a gazelle, and as smart as a person, while remaining as able to thrive amid squalor as a cockroach, but there is no chance that such a string of happy accidents could occur.

There are many different "constraints on perfection." Some of them have been identified by one of the most challenging and vocal of sociobiologists, Richard Dawkins, in his recent book, *The Extended Phenotype*.

Constraints on Perfection

NEUTRAL CHARACTERISTICS.

Not all of an animal's characteristics are the product of natural selection. Some characteristics have neither positive nor negative selective value, but ride on the backs of other characteristics that are adaptive. Furthermore, once there, these neutral characteristics may set constraints on what further modifications of the organism will be improvements. The existence of "neutral" characteristics results in part from the fact that genes rarely have only a single effect. If some of the gene's effects are positive, and thus selected for, the others will be carried along. In short, there can be evolution without natural selection.

To clarify this point, consider a common toy for young children, a clear, plastic cylinder that is separated into three tiers. Each tier has holes in its floor, and the holes are of different diameter, with the diameter getting smaller as one moves down the cylinder. If a bunch of different sized balls were put in the top of the cylinder, and shaken, the cylinder would end up with the different sized balls each occupying a different tier, with the largest on the top and the smallest on the bottom. This is because the diameter of the holes serves to select balls of a given size. Now suppose that all the biggest balls were red, all the medium-sized balls were blue, and all the smallest balls were green. Shaking the balls would separate them not only by size, but by color. But to look at all the green balls on the bottom and say that green was "selected for" would be a mistake. What was selected *for* was size, not color. What was actually selected was both size and color. Philosopher Elliot Sober, in his book, *The Nature of Selection,* argues that it is critical to distinguish between "selection for" and "selection of." To confuse these two ideas is to treat every characteristic observed in animals as the direct result of selection. But some characteristics of animals, like the color of the balls, are not the result of selection. They are only along for the ride.

TIME LAGS

There are serious time lags in evolution. Behavior may be nonmaximizing now, but it may have been optimal under the conditions in which it evolved. It is environmental conditions, after all, that determine the adaptiveness of any characteristic, and environmental conditions change. Since evolution typically works slowly, it will take time for evolutionary change to catch up with environmental change. Thus, for example, there is a particular species of bird that lays only one egg at a time. It has been shown experimentally that these birds are perfectly capable of supporting and nourishing two offspring if a second egg is added to the nest. Thus, the reproductive behavior of this species is significantly nonoptimal. However, it is quite possible that at the time that this reproductive pattern developed, the environment offered much less potential food, so that two offspring could not have been successfully nurtured.

This time lag is especially significant in the case of human behavior. People, through language and culture, modify their environments at rates that evolution could not possibly match. In industrial societies, people live in environments that are almost wholly artificial. It is hardly surprising that *natural* selection yields traits that are not optimal in *artificial* environments. Indeed, the human environment is almost unrecognizably different now from what it was a few hundred years ago. But a few hundred years is not even a second of evolutionary time. Thus, when we look at many current human social practices that seem not to be in the interest of inclusive fitness maximization, it must be remembered how different the world was when they first developed.

CONTEXT DEPENDENCY

Evolution of a characteristic always occurs in the context of other characteristics that an organism might possess. Homosexuality certainly does not seem to be an inclusive fitness maximizing way to behave. So how can it be a biologically determined trait? The answer may be that when the genes for homosexuality first evolved they were not genes for homosexuality. They may have contributed to the determination of other traits, or they may have had no behavioral effect at all. Further evolutionary change in the organism may have turned these genes into genes for homosexuality. Said another way, there is many a slip between genotype (genetic constitution) and phenotype (behavioral manifestation).

HISTORICAL CONSTRAINTS

History constrains adaptation. When an organism changes in major ways during evolution, it doesn't do so all at once. It does so in a long series of stages, each of them only slightly different from the one preceding it. For this gradual process to continue—for the organism to get from trait A to trait Z, all the intermediate stages (traits B, C, D, . . .) must themselves confer some selective advantage. Trait Z′ might be better than trait Z, but the organism just can't get there from here. The best end point is not the best one possible, but the best one that the can be arrived at through a series of stages each of which is an improvement on what preceded it in the environment in which it occurred.

To use Dawkins' own example, the jet airplane engine is the technological marvel it is partly because it was designed from scratch, unconstrained by the propellers that preceded it. But suppose that jet engine designers were required to make a jet engine out of a propeller engine. Suppose further that each modification in the propeller engine en route to its becoming a jet engine had to be an improvement on its predecessor. With these constraints, people may never have got to a jet engine at all, let alone one that works effectively. But these are precisely the constraints within which evolution must work.

VARIATION CONSTRAINT

Natural selection only works on what occurs, and thus maximization is constrained by the limits of actual variation displayed by members of a species. In part, this is a matter of timing. A given variation may occur, but at an evolutionary time when it is not especially adaptive. If so, it won't be selected. So organisms need to make the right moves at the right times (or rather, their genes need to). Indeed, as discussed earlier, the importance of variation to the evolutionary process has been pointed to as an explanation of why reproduction is so often sexual rather than asexual.

COST–BENEFIT ANALYSIS

All adaptations have costs. Evolving the best possible defense against predation (say, a coat of armor) may make an animal so immobile that it can't catch any food. All adaptations are the result of compromises produced by evolution's cost–benefit analysis. Biologists may treat feeding, sex, caring for young, and defense against predation as isolated functional systems, for purposes of analysis. And they may design an optimal strategy for accomplishing each function. But it must be remembered that in real life these systems coexist in individual organisms. Invulnerability to attack doesn't do much good if its by-product is starvation.

LEVELS OF ANALYSIS

How good an adaptation is may depend on what level of organization one is looking at. An adaptation may be great for the genes involved, but less great for the organism that houses those genes, and even worse for the social group that houses those individuals.

CAPRICIOUS ENVIRONMENTS

Finally, organisms are faced with an unpredictable environment. An organism may slowly, painstakingly evolve bit by bit some characteristic that makes it perfectly suited to its environment, and just as it is about to enjoy the fruits of all this evolving, the environment can change dramatically. A sudden change in environmental conditions may upset the work of hundreds of thousands of years of gradual evolution.

With all these forces working against perfectibility, it's a wonder that organisms are able to survive at all, let alone act in the most effective possible fashion. Thus, while biologists may talk about maximization as a theoretical abstraction, it must be understood that what they really mean is maximization subject to constraints, and that the constraints are sufficiently forbidding that traits approaching actual maximization virtually never occur.

This same conclusion was reached in the discussion of rational economic individuals. People never adopt the best imaginable strategy; they adopt the best one subject to constraints. Constraints are imposed by available income, by available commodities, by the time required to make economic decisions and act on them, by the time required to gather information about possible alternative actions, and so on. Indeed, the constraints on economic maximization are so forbidding that it is more accurate to think of people as "satisficing" than to think of them as maximizing. The "satisficing" person seeks a satisfactory outcome, not the best one possible.

Where does all this qualification on maximization leave sociobiology? In one respect, it is sociobiology's salvation. For if organisms really did maximize, we would have to rewrite evolutionary theory. A world of maximizing organisms would indeed be the best of all possible worlds. No improvement would be psoible. Evolution would have reached its terminus. But how could a series of random variations acted upon by accidental environmental circumstances result in perfection of all things? It couldn't. An evolutionary story like this one would constitute a return to the divine creationism it was intended to replace. It would become teleological, with organisms evolving always toward a known *telos*, or goal, and then, on reaching it, staying put. It would change the operative evolutionary principle from "the one better" to "the one best." But what evolutionary theory implies is that no one can know what the "one best" will be. How unlikely that all species could arrive at it! Only God can make a perfect tree.

Thus, an acknowledgment of constraints on perfection by sociobiologists is critical because without it they would no longer be evolutionists, as we and they understand it. But this faithful bow in the direction of evolutionary theory has substantial consequences. If it is to be more than lip service, if we are to take constraints seriously and thus take maximization away from sociobiology, precious little is left.

Think about "maximization subject to constraint," with constraints meant to indicate some of the items on the list just reviewed. What are the rules for deciding what is to count as a maximizing trait and what is to count as a background constraint? The salmon swims upstream, lays its eggs, and dies. If the critical trait is the thousands of eggs the salmon lays and the contortions its innards go through to make that possible, and if the background constraints are the arduous trip upstream and the fact that it spawns only once, then the salmon's behavior seems to fit with a constrained maximization theory. But why treat the trip upstream and one-shot spawning as background? Why not treat the fact that it lays thousands of eggs as background, and evaluate the utility of destroying its insides and dying? This hardly seems like an example of maximization. One would expect a mutant salmon able to lay all those eggs without dying to spread its genes like wildfire through the population. The current specimen seems to be making rather a mess of things.

The point, of course, is not to think of the salmon as ill adapted, or to propose "survival of the misfits" as an alternative to Darwinian theory. Rather, the point is that the notion of maximization subject to constraint says little unless it can provide a set of principles for treating some characteristics as constraints and others as maximizing traits. Any activity at all can be shown to be maximizing fitness if the constraints are drawn carefully enough. Indeed, a given trait and its opposite can both be seen as maximizing fitness with constraints suitably drawn. Theories with this much flexibility can explain everything; as a result, they explain nothing.

The strategy that sociobiologists adopt in attempting to fit the remarkable diversity of animal behavior patterns into a unifying maximization framework is in essence to assume as true the very principle under scrutiny. "We know", they say, "that organisms maximize inclusive fitness subject to constraints. So when we look at a bit of behavior, let us ask what the constraints must be such that this behavior is the best the animal can possibly do." With no rules about what is to count as a constraint, it will always be possible to tell a tale about how a given behavior is fitness maximizing. Such tales are what Stephen Gould and Richard Lewontin have termed "just so" stories. The salmon spawns only once, killing itself in the process. How is this fitness maximizing? Well, it must be that one-time spawning is a background constraint. Many birds form stable mating pairs, with the males devoting considerable energy to the raising of their young.

This may seem like a waste of good sperm, counter to the male promiscuity strategy that sociobiological analysis would otherwise predict. If so, it must be that these birds are constrained biologically to pair off and that they do the best they can subject to that constraint.

So pair bonding is fitness maximizing where it occurs; promiscuity is fitness maximizing where it occurs. Multiple spawnings are fitness maximizing, as are single spawnings. Brute selfishness maximizes fitness, but so does self-sacrifice. Menopause maximizes inclusive fitness because at a certain age females can do better for their genes by being helpful grandparents than by producing offspring of their own. Why? Because menopause wouldn't be around if it weren't fitness maximizing. But fertility throughout the life cycle is also fitness maximizing. Even death is fitness maximizing, since it allows an old animal's potential share of scarce resources to go to younger and more vigorous genetic relatives. With "background constraints on perfection" operating as a theoretical wild card, there is no possible finding—no pattern of animal behavior in nature—that can embarrass sociobiology.

An example of this ability to produce explanations on demand can be found in some sociobiological analyses of human sexuality. Sociobiological analysis predicts that, in general, females, not males, will be selective in choosing mates. This is principally because the investment of the female in mating is much greater than that of the male. Because mating costs her so much, she wants to be sure to do it right. Female selectivity might naturally suggest that by and large males will be trying to make themselves attractive to females. And frequently in nature, this is indeed the case. It is the male, not the female, in most species of birds, whose appearance is brightly colored and heavily adorned and whose mating song is long, loud, and distinctive.

Obviously (at least in our culture), this analysis does not apply very well to humans. With humans, it is the females who make themselves attractive, painting their faces and their fingernails; the males are drab. Indeed, it would seem that with people, it is the males who are doing the selecting. A problem for sociobiology? Not at all. According to Donald Symons, in *The Evolution of Human Sexuality,* this pattern is just what should be expected. Large breasts and wide hips advertise the female's reproductive fitness. And male drabness is an advertisement of conservativeness, an indication that the male is likely to be steadfast and a good provider. It's the males who walk around with shirts unbuttoned to their navels, and gold chains dangling from their necks who are likely to be promiscuous and irresponsible. So females actually select drab males. The sexual attractiveness of women is their device for controlling physically more powerful men. "In the West," says Symons, "as in all human societies, copulation is usually a female service or favor."

If sociobiology has so little explanatory power, one must wonder, why is it taken seriously? Perhaps it is because sociobiology borrows its explanatory framework from another context in which such explanations *are* plausible. Imagine being in the house of a stranger. It is an unusual house. Everything is on a single level. A ramp, not a stairway, leads to the front door. Entrances to rooms are unusually wide and are closed off by curtains rather than doors. All shelves and cabinets are waist high or lower. So are the light switches on the walls. Even the sinks and stove are so low that an adult would have to bend or sit to use them. The owner of the house rides around all the time in a motorized chair with wheels. He never gets up from the chair. The chair fits through all the doors easily. He can go from room to room just by brushing aside the curtains. The shelves, sinks, stove, and so on are just the perfect height for him to have access to them without ever leaving his chair. Though it is rather unsuited to most people, the house seems perfectly designed for its owner.

Why would anyone choose to live in so peculiar a fashion? The answer, of course, is that the owner made his choices subject to a background constraint: he was paralyzed and confined to a wheelchair. And how did this perfect, utility-maximizing fit between organism and environment come about? Did the man find himself in this house and decide to be paralyzed to take advantage of it? Did he wake up one day paralyzed and go stumbling about at random until he happened on an environment that served his paralyzed state well? Or did he design an environment—with goals, purposes, and intentions clearly in mind—that would cater to his needs given his constraints? The answer is clear; the choices are even absurd. The seemingly perfect fit between man and house is the result of his analysis of his problems, his assessment of various possible solutions, his assessment of the limits of his resources, physical and financial, and his appreciation of his goals. It is clear what the background constraints on his utility maximization were. He was paralyzed first; everything else came as a result of his paralysis. And the everything else that came was the result of his direct and directing intervention.

Of the three possible accounts of the observed fit between the paralyzed man and his environment, the only one that is reasonable is precisely the one that the sociobiologist cannot appeal to. It is teleological; evolution is not. It has the man shaping his environment. Evolution tells us that environments shape organisms (by affording selection pressures of various kinds). The accurate evolutionary kind of account sees the fit as the product of random stumbling about until the right environment is encountered. The poverty of sociobiology is reflected in the fact that it cannot distinguish becoming paralyzed to adapt to one's house from building one's house to adapt to being paralyzed. Which of these stories is the right one in any given case will depend upon how one construes the background constraints on perfection. And sociobiology has no rules for construing background constraints.

There are two ways to make sense of the fit between the man and his house. First, knowing nothing about the man, one just assumes that he is rational and wouldn't do anything so crazy as to paralyze himself. And knowing something about the world, one infers that houses as peculiar as this do not come about by accident. Second, instead of resorting to inference, one can find out about the man's history, by determining empirically that the paralysis preceded the house. Once it is known that paralysis came first, it is possible to talk with justification about what constrained what. Paralysis didn't determine the building of the house, but it certainly had an influence on which kinds of environments would be more satisfactory than others.

Evolutionary biology, properly understood, has only one of these two options open to it. It can't talk about the "irrationality" of this or that behavior pattern. Nature isn't rational; the natural world has not been planned. It can, however, appeal to natural history. The one legitimate way to constrain talk about constraints is to know what came first in the natural evolutionary history of a species. An organism's collective attributes at a given point in evolutionary time establish constraints on what it can become in the immediate future. A salmon can't become a shark overnight. Knowing an organism's natural history allows guesses at what kinds of variations in behavior were possible at any given time. Knowing about the environment in which the animal lived allows guesses at which of these variations might have had favorable consequences for the reproductive success of the species. In this way, plausible historical narratives can be constructed about why a species evolved along one set of lines rather than another.

Note that this kind of historical narrative offers little in support of the idea of "maximization subject to constraint." Even if it were known that at a certain time the salmon was already condemned to a single spawning in its life, it would remain mysterious why the salmon "maximizes" by turning its insides out to lay so many eggs rather than by evolving the ability to spawn more than once. Said another way, even if everything about a species' past is known, one cannot appeal to maximization of fitness, or any other principle, to predict the species' future. The idea of fitness maximization can be used to rule out lots of possible futures, like one in which the salmon spawns once, lays only a few eggs, and dies, but when all the ruling out is done, there will still be a countlessly large number of reasonable possibilities left. When the next change in a species occurs, it can sensibly be incorporated into the historical narrative. But it is not possible to unite various historical narratives under a single, universal, natural law.

With this view of evolutionary theory as a kind of historical narrative, consider the question: What does it mean to be a cockroach? What is the cockroach's essential nature? It is possible to say what it means to be a cockroach now, by pointing out whatever characteristics systematic biologists use to distinguish cockroaches from other species. But what it means to be a cockroach now may not be what it always meant, eons ago, in the cockroach's evolutionary past, nor what it will mean in the cockroach's evolutionary future. About the past, one can find things out, by learning of the natural history of the cockroach and its relatives. It might be possible to tell an interesting story about how "essence of cockroach" has changed. But the future is largely uncertain.

If "essence of cockroach" does not lie out there in the world, true for all times, to be sniffed out by scientists the way pigs hunt truffles, what about "essence of human nature"—rational economic man as articulated by economists and naturalized by sociobiologists. The very essence of one line of criticism of classical economics discussed in the last chapter is that "rational, self-interested, economic man," if it is a true account of human nature at all, is true only at a particular time and place. "No misreading of the past ever proved more prophetic of the future," said Karl Polanyi of Adam Smith. "Essence of person," like "essence of cockroach," changes—not randomly or capriciously, but subject to constraints imposed by the past as well as opportunities provided by the present. Thus, explaining the pursuit of self-interest by appealing to "human nature" is no explanation at all, and even less is it a justification. How did human nature come to be this way? How can it be changed? How should it be changed? These are all questions that an evolutionary view of the nature of organisms does nothing to silence—and little to answer.

Evolutionary Biology and Human Nature

These questions lead to a second line of criticism of evolutionary biology. Some critics are willing to bypass the arguments of the last section and concede to sociobiology its power to account for much of the behavior of animals. But its power stops, they say, at humans. There are two reasons for this. First, people are rational. They articulate goals, formulate plans for achieving those goals, and act on those plans. They are not slaves to the opportunism of chance variation and selection. They make variations happen, and they alter environments in accord with their designs. Paralyzed men *build* suitable houses; they don't stumble onto them. Natural selection may have taken teleology and design out of nature, but people are teleological, designing creatures.

Second, with the emergence of human beings, there came a new influence on action that does not exist in the rest of nature. That influence is culture. People live in cultures, and these cultures have an enormous, nonbiological influence on what they do. Culture provides an alternative to the genes as a route of transmission of behavioral characteristics from generation to generation. And culture-induced changes in behavioral characteristics progress much, much faster than genetic ones. Culture has, in effect, made evolution obsolete. Thus human rationality and human culture pose challenges to the sociobiological explanation of human nature.

Biology and Rationality

Natural selection provides an unintelligent, nonteleological mechanism to account for what seems to be highly intelligent and goal-directed characteristics of organisms in the natural world. Fish have gills, birds have wings, and giraffes have long necks not because either God or their ancestors figures out that these would be useful features to have. These features—the usuful ones—are the ones that survived the process of variation, competition, and selection. Organisms possessing other features that were not so useful have left no descendants.

It may be quite reasonable to understand evolution from within this framework, but attempts to apply it to the lives of individual people can be quite ludicrous, as in the case of the paralyzed man. While one could tell a tale about how the fit between the man and his environment was the successful product of random variation and selection, such a tale would have little to do with the truth. The paralyzed man, in designing his environment, is the paradigm of rationality. He formulates goals, anticipates difficulties, and then builds a world in which these difficulties can be overcome. His behavior is about as far from random variation and selection as one can get.

It may be argued that even if natural selection is not rational (goal-directed, planful, and so on), its results are. That is, the products of natural selection—organisms well-adapted to their environments—are the same as the products of rationality would be. Natural selection serves the function that rationality would serve; it makes nature look as if it were rational. It makes cockroaches act as if they were rational. Perhaps, it also makes people act *as if* they were rational. This possibility is worth exploring. Doing so requires a comparison of natural selection with the everyday meaning of rationality.

Think for a moment about how natural selection produces intelligent results. To aid discussion, focus concretely on how a species might evolve in the direction of ever increasing running speed. How might foxes become fast enough to catch rabbits? For such a development to occur, two ingredients are required. First, the raw materials must be present. Variation must produce occasional foxes that run faster then their peers. Second, when the variation occurs, it must make a difference. If even slow foxes catch all the rabbits they need, or if even the fast foxes fail to keep up with rabbits, there is no reason to expect that foxes will evolve in the direction of increased running speed. Remember, evolution occurs not because anyone figures out that this or that trait *might*, someday, be helpful. It occurs because a trait has an immediate effect on reproductive fitness. If even slow foxes catch rabbits, they will not be at a reproductive disadvantage compared to fast foxes. Therefore, there will be no selection pressure pushing foxes in the direction of increased speed. Similarly, if the fast foxes are still not fast enough, their speed will not give them a reproductive advantage over slow ones. Recall that one of the constraints on perfection identified by Dawkins is that intermediate evolutionary forms must themselves confer some selective advantage. If B is not better than A, the animal will never get to Z. So foxes as a species will never get very fast unless somewhere along the line foxes that are slightly faster than their brothers are able to catch rabbits that their brothers can't catch.

Now imagine a rational fox. How would its rationality make a difference? First of all, it would not have to sit and wait for a lucky variational accident. It could make the variation happen. "Suppose I could run faster," it might say to itself. Second, if the fox were having no trouble catching rabbits at its current speed, it could still start running faster. "I may not need it now," it might say, "but rabbits could get faster, or smarter, or they might become scarce, in which case I'll have an edge over other foxes." Or alternatively, if the fox gets faster, but still isn't fast enough, it might say: "I'm getting closer to those rabbits all the time. I'm clearly moving in the right direction. A little more practice, and I'll be there." Or it might say: "This effort to run fast is silly. I'll never be able to catch up with those rabbits. I ought to try something different. Maybe I should work on outsmarting them. That's it; I'll work on being 'foxy' instead of fast." Operating with true rationality, each of these moves is possible. Operating with the evolutionary simulation of rationality, none of them is possible.

Evolution is a fundamentally opportunistic process. A chance variation occurs, and the species says yes or no to it right then and there. It says yes if the variation yields increased reproductive fitness; otherwise it says no. It can't say yes to a variation that looks promising for the future. It can't say no to a favorable variation in the hope or expectation that a still more favorable one will occur later. Opportunism is a form of myopia. Evolution walks forward, but it is always looking straight down at its feet.

That evolution is opportunistic means that it can't pass up improvements, no matter how small they may be. Furthermore, while maximization of fitness may be the engine of evolution in the mind of the evolutionary biologist, it is not in the mind of the evolving organism. There is no overall fitness maximum. Evolution knows only about better, not about best. One reason for this is that what in the future might be best will depend upon what the species is actually like. But what the species is actually like will in turn depend on which chance variations it has already said yes to. For each time the species says yes to a variation, it changes. And one consequence of the change is that the criteria for saying yes in the future will change. That is why one can't predict the course that evolution will take. There may be only one "best," but there are lots and lots of "betters," and which will actually occur is a matter of luck.

True rationality, unlike evolution, need not be opportunistic. As

Jon Elster points out in his book, *Ulysses and the Sirens*, rational actors are capable of goal-directed strategies that evolution can't make use of. For example, rational actors are capable of waiting. They can pass up opportunities for small gains in the light of prospects for larger ones. They can wait to buy a stock whose price is falling if they think it will fall still further, and they can wait to sell a stock whose price is rising if they think it will continue to rise.

In addition to being able to wait, rational actors are able to progress toward goals *indirectly*. They are able to take several steps backward if they can foresee that such a move will permit even more steps forward later on. The good chess player looks many moves ahead and is quite willing to take short-term losses that will eventuate in a winning position. The good bridge player intentionally loses tricks he could win early in the play of a hand, to set up a favorable position at the end of the hand.

There is another significant respect in which rationality differs from evolution. It concerns the process of variation. We have already seen that while evolution must wait for fortunate, chance variations to occur, people have the capacity to make variations. The rational fox can make himself run faster, while the nonrational one awaits happy accidents. But just as important as the capacity to make variations is that people can make variations that are *functionally relevant*. The source of variation that evolution works on is the genes. Genes don't know about running, or about flying, or about long necks. Genes are just biochemicals, and there is no reason to expect that the units of biochemical variation will in general be directly relevant to the functioning of the organism.

To use an example of Elster's, it is possible to increase the likelihood that a book will contain misprints (analogous to increasing variation) by breaking the eyeglasses of the typesetter. However, doing this will not increase the likelihood of misprints in the second edition of a book that correct the mistakes in the first edition. Rational actors who make variations are able to make variations that are functionally relevant. In evolution, the mechanism for producing variation and the mechanism for selection are independent of each other. In rational action, they are not. People make variations with the goal of those variations very much in mind.

The sociobiologist might offer a number of responses to the suggestion that human rationality limits the relevance of sociobiological accounts of human behavior. First, the sociobiologist might agree. He might suggest that the sociobiological framework can account for those aspects of human behavior that rationality doesn't reach. Or he might suggest that rationality modulates fitness-maximizing tendencies of precisely the sort uncovered in nonhumans. In nonhumans, natural selection has determined actual *behaviors;* in humans, it has only determined *tendencies,* which can be overridden. Or he might suggest that humans *use* rationality to serve precisely the same fitness-maximizing ends that are served in other animals by other means.

However, to take any of these positions is to limit severely the relevance of sociobiology to human life. Who knows where rationality can reach? Who is to say that what is outside its compass today will still be outside it tomorrow? Behavior that is as central to sociobiological concerns as reproduction may have been outside the bounds of rationality five hundred years ago, but is certainly well within its scope now. People plan the size of their families and the spacing of their offspring. It is now within our power to change the ratio of the sexes, by means of chromosomal examination of the fetus and selective abortion. The sociobiologist who concedes this power to rationality concedes a great deal.

Alternatively, the sociobiologist may want to suggest that this appeal to rationality is overblown. People more often than not do behave opportunistically. They do pursue short-term interests without much of an eye toward long-term consequences. Men and women court and mate in close accord with sociobiological predictions. Even if it is true that what are determined behaviors for nonhumnas are only tendencies for humans, they are powerful tendencies—ones that are only rarely violated or supervened by other influences. Furthermore, when people say that they are doing such and such in order to achieve so and so, why should they be taken at their word? Why assume that people know what they want and why they behave as they do? Why grant to people that they are rational just because they say they are? After all, people used to see rationality throughout the natural world, before Darwin. And much of what is casually identified as rational can easily be redescribed with the concepts of variation and selection. The entire appeal to rationality may be a massive act of self-deception, one whose adaptive significance would have to be determined.

To suggest an account like this, the sociobiologist would have to offer a plausible alternative explanation of actions that certainly appear on the face of it to be rational. A handy framework for such an alternative is available. It is sometimes referred to as a *functional explanation.* The general character of functional explanations is to show that some phenomenon that *could* have been produced by people acting with complete rationality, control, and knowledge of the situation was instead produced inadvertently, as a consequence of the functions it served for the individuals involved. The theory of natural selection is the hallmark example of functional explanation. Species changes that could have been the product of intelligent design are instead shown to be the product of unintelligent variation and selection.

There have been numerous attempts to apply the logic of functional explanation to human, social phenomena. One of them is Adam Smith's notion of the "invisible hand." Remember that, according to Smith, a society full of egoists, each concerned only with his own self-interest, if allowed to exchange in a free market, will produce social and economic results that are good for all. Not only good for all, but at least as good as any planned, concerted social action could produce. The butcher and the baker neither know nor care how their individual actions will affect society. They are not concerned with promoting social welfare. Nevertheless, promote social welfare is what they do, if they pursue self-interest in the right (free-market) circumstances.

Functional stories like this have plausibility. There is little doubt that as descriptions of the phenomena under scrutiny they are quite reasonable. That is, one could *describe* the market as serving the function of providing for the collective welfare. Similarly, one could *describe* the adaptive characteristics of animals as the result of natural selection. The point of functional analysis, however, is not just to offer *descriptions* but to offer *explanations*. Yes, the functionalist argues, God could have designed adaptive creatures, but he didn't. Evolution did.

Choosing between rational and functional explanations is often difficult, but is has important implications. It is difficult because sometimes two examples of the same phenomenon can exist side by side, in the same person, at the same time, and one of them warrants a rational explanation while the other warrants a functional one. Consider a space engineer who plays the outfield on the company softball team. One day at work, he figures out where in the Pacific Ocean a satellite is going to land, so that a ship can be dispatched to pick it up. Then after work, he plays in a game and catches several fly balls. Catching a fly ball in a game and knowing where a satellite will land both involve knowledge of the physics of projectile motion. But while the engineer uses that knowledge on the job, he uses functional rules-of-thumb instead in the outfield. If he got confused and started using rules-of-thumb on the job and physics equations in the outfield, he would be out of work and out of a position on the team.

Or, to get closer to the heart of sociobiology, consider the notion of kin selection—of inclusive reproductive fitness. Sociobiology argues that organisms will take risks for their kin, and how big a risk they will take depends upon how many kin are involved and how closely related they are. It is worthwhile to take a 50 percent risk to save three children, but not to save three cousins. Now the sociobiologist wouldn't dream of suggesting that birds, fish, and worms take out their calculators and perform the appropriate calculations before deciding whether to take the risk or not. Rather, whatever mechanism determines risk taking is sensitive to the degree of relatedness of the organisms involved. Otherwise, it would not have been selected in evolution. So this mechanism has the effect, or function, of serving inclusive fitness, though the organism may know nothing about it.

On the other hand, the organism may know about it. When people take risks for their kin, they may be uncalculated, nonrational, unintended. But they may be the product of rational deliberation. Parents who build up financial assets to be used by their children may know precisely what they are doing and why they are doing it. In the human domain, both functional and rational explanations of risk taking or sacrifice for relatives are serious candidates for allegiance. The mere fact that people do sacrifice for their young, just as birds do, does not mean that the mechanisms involved are the same, or even approximately similar, in the two cases. And if the mechanisms are not similar, then the only relation between the sociobiological analysis of the care of young by birds and the phenomena of the care of young by people is one of weak, though perhaps suggestive, analogy.

Functionalism, Rationality, and Cooperation

A particular area of sociobiological explanation where the distinction between functional and rational explanations is extremely important is in the analysis of competition and cooperation, or of aggression and altruism. Suppose that it is in the best interests of a group for its members to be cooperative and altruistic toward each other—that groups whose members share food and work collectively to rear young and repel enemies will be better off than groups of completely selfish individuals. How can evolution work to produce such cooperative groups?

By and large, sociobiologists do not regard the idea of group selection as viable. Even if a cooperative group is in better shape than a competitive one, natural selection doesn't operate at the level of groups. Selection operates at the level of individuals, or perhaps, at the level of genes. So cooperation must be in the interests of each individual's genes. An account of how cooperation is in the interests of each individual's genes is possible with the concepts of inclusive fitness and kin selection. If by cooperating I promote the survival and reproduction of many relatives who share my genes, there will be more of those genes in the next generation. This is how selfish genes can yield self-sacrificing individuals.

However, a population full of self-sacrificing individuals will not be stable. If a mutant individual should appear who is brutally selfish, that individual can reap all the benefits of a cooperative group with none of the attendant costs. As a result, that individual's "free-rider" genes will spread through the population, gradually transforming it into a collection of ruthlessly selfish individuals. But that will also not be stable. Once everyone is selfish, small groups of individuals who cooperate will have an advantage over the selfish ones. An evolutionary stable strategy (ESS), which is a strategy that cannot be improved upon by any other that is adopted by the bulk of a population, will be some mix of cooperation and selfishness, with the particular proportions of each determined by the relative costs and benefits of being selfish or cooperative. Such a mix will not be optimal—full cooperation would probably be better—but it will be good enough, and it will be stable.

Now one of the main points of such a sociobiological analysis is to show how it is possible to give a functional explanation of a pattern of group social behavior that could also be described in rational or intentional language. That is, one could imagine setting out to design rules for social interaction that would be effective for the members of a social group. The concept of an evolutionary stable strategy suggests that such rules can emerge even without a designer. It is tempting to suppose that the rules of human social interaction that determine patterns of cooperation and aggression in human societies have evolved in just this nonrational way. Such rules would not make for the best of all possible worlds, but they would make for worlds that are good enough and are stable.

How can one decide whether such a functional account of patterns of cooperation and selfishness is correct? To earn allegiance, a functional account must have at least two characteristics. First, it must yield results that are stable, that are invulnerable to invasion by other patterns of social interaction. But second, and just as important, it must yield results that are *individually accessible*. To see what this means, imagine a group composed of only ruthlessly selfish individuals. This sort of group is not stable. A subgroup of cooperators will do better. But how can that subgroup form? Individual animals cannot plot and plan and enter into alliances for the common good. Each individual is on its own. So suppose a cooperator is born into a selfish group. Will it succeed? Of course not. It will be swallowed up by its thoroughly exploitative fellow group members before it has a chance to reproduce more of its kind. By the time a second cooperator happens, by chance, to be born, the first one will be long gone. For ultimately advantageous cooperation to get started, there must be the very happy accident that several cooperators are born at the same time, can recognize each other, and form a cooperative subgroup. Thus, it is not clear that a cooperative component of group interaction, even if stable, is accessible via the independent action of individuals.

A great deal of light has been shed on this issue of cooperation versus competition by psychologists, economists, and political scientists, among others, through the study of a peculiar kind of game. The game is called the *prisoner's dilemma*. Unlike most of the games with which people are familiar—games like chess, football, poker, and so on—in the prisoner's dilemma, both players can win and both players can lose. They are not necessarily playing *against* each other. Most everyday games are known as "zero-sum" games; the amount won will always equal the amount lost. What one person wins is directly at the other person's expense. The prisoner's dilemma is a non-zero-sum game. Both players can win, opening up the possibility for cooperative action.

The game derives its name from the following hypothetical situation. Two people suspected of commiting a major crime are arrested and interrogated separately. The idea behind the interrogation is to induce a confession. Each is told that if he alone confesses, he will be granted immunity in return for being a witness against his partner, while his partner will get ten years in jail. However, if both confess, they will each get a jail sentence of five years (one doesn't need testimony when one has a confession in hand). The penalty is only five years instead of ten because their willingness to cooperate serves to mitigate the crime. Finally, if neither one confesses, they will each receive one year in jail, as punishment for a lesser offense for which the authorities have independent evidence.

What should a prisoner do? Should he cooperate with his fellow prisoner and stay silent, or should he defect and confess? Let's examine what is in it for him. Suppose his partner cooperates and stays silent. In that case, he will get one year for also staying silent, but he will go free for defecting. So it is in his best interests to defect. And suppose his partner defects. Now, he will get ten years for staying silent, and only five years for defecting. So again, it is in his best interests to defect. In other words, no matter what his partner decides to do, he is better off defecting.

What about his partner? Exactly the same considerations apply to him. He is better off defecting, no matter what the other prisoner does. So both prisoners, being rational pursuers of self-interest, will defect. The result is that each will get five years in prison, instead of the one year they would have gotten if they had cooperated and stayed silent. If only the prisoners had reached agreement beforehand, they would both have ended up better off, with one year in prison instead of five. In the absence of an agreement, each prisoner acting independently was compelled to make a choice that made them both worse off. But would an agreement have helped? Suppose they had agreed to be silent, and then, sitting alone in that room about to be interrogated, a horrible thought occurred to one of them: "Suppose my partner violates the agreement. If he trusts me to be honorable, he can go free by confessing. And I'll get ten years. He'll never honor our agreement. I better confess." And of course the other one, at the same moment, has the same horrible thought and makes the same unfortunate decision. There seems to be no way out of it.

The prisoner's dilemma needn't involve jail sentences and confessions. Indeed, one of the reasons that it has been studied so extensively is that it seems like a good model for many different social situations that involve choices between cooperation and defection or competition. The seemingly never-ending arms race between the Soviet Union and the United States fits the prisoner's dilemma rather well. Both nations stand to gain from mutual cooperation (disarmament). However, both are concerned about the risks involved (suppose we sign a treaty and then we honor it and they don't). The result is that both nations waste unimaginably vast resources on armaments and military training.

The prisoner's dilemma also operates at the level of individuals. If individual workers could agree to work only so hard and no harder, they could ease a potential source of pressure from their employers to get them to compete with one another. But once such an agreement has been reached, an individual worker can leapfrog over his fellows by violating it and working very hard. Similarly, students could force their teachers to modify expectations about classroom performance by agreeing to study only moderately hard. And world-class athletes could free themselves from the potentially disastrous long-term effects of performance-enhancing drugs if they agreed collectively not to use them. The examples proliferate endlessly.

In general, the prisoner's dilemma requires the following characteristics. First, the best a person can do is defect while the other person cooperates. Second, the worst a person can do is cooperate while the other person defects. Third, mutual cooperation is better than mutual defection, with both intermediate between one person cooperating and the other defecting. Thus, if we use C as a symbol for cooperating, and D as a symbol for defecting, Person A's preferences in a prisoner's dilemma game with Person B would be thus:

	A	B
1	D	C
2	C	C
3	D	D
4	C	D

Outcome 1 must be better for A than 2, or else A has no temptation to defect. Outcome 2 must be better than 3, or else A has no temptation to cooperate.

Suppose that a prisoner's dilemma game involved lots of moves, lots of turns. Indeed, suppose that the game was never-ending, that it just went on indefinitely. Is there any way that the two players, without communication or collusion, could end up better than just choosing to defect each time? Under what conditions, if any, can the dilemma be satisfactorily resolved? Is there an optimal way to behave in a situation like this? Is there a way to ensure mutual cooperation? Does such cooperation demand rationality, social coordination, honor, and altruism, or can it be developed by nonrational, egoistic, dishonest individuals—by selfish genes in selfish bodies?

These questions have recently been addressed in a book by Robert Axelrod, called *The Evolution of Cooperation*. What Axelrod did was organize a prisoner's dilemma tournament. He invited people to develop strategies for playing the game, and then he pitted those strategies against one another. The people who invented these strategies didn't

actually play against each other. Instead, a computer was pro-
grammed to follow the strategies, and it simulated thousands of turns
with each strategy pitted against the others. The winner would be
the strategy that survived the competition with the most points.

And the winner was a surprise, first, because it was very simple,
and second, because it was nice. Even in a situation that was explicitly constructed to get people to try and maximize their own interests,
without regard for their fellows, a nice, cooperative strategy did better than any competitive ones. The strategy can be called TIT FOR
TAT. What it specified was to start out cooperating on the first turn
and from then on to do whatever the opponent had done on the previous turn. Thus, it rewarded an opponent's cooperation by cooperating itself, and it punished an opponent's defection by defecting. But
it wasn't vengeful; it didn't keep defecting. If the opponent "saw the
light" and cooperated on a turn, TIT FOR TAT immediately switched
to cooperation itself. The conclusion Axelrod drew from this contest
was that it was possible for cooperation to emerge as the dominant
strategy even in a world full of self-interested egoists who could not
coerce others to cooperate, enter into explicit alliances with them, or
even communicate with them. A world of selfish genes did not necessarily have to end up as a world of callously competitive individuals.

Axelrod identified several conditions that had to be met in order
for a nice strategy to be successful:

1. *The future has to matter.* If all one cares about is the outcome of
any one turn, he does best by defecting. It is only the potential bad
future consequences of defecting (on later turns) that can induce
cooperation.

2. *One's opponent must be responsive to one's own choices.* If the
opponent is choosing at random, being nice will do you no good in the
future.

3. *Furthermore, the chances must be good that opponents will encounter
each other again in the future.* Opponents have to recognize each other
and remember how they acted in the past. If opponents didn't know
that they had met before, or didn't remember what had happened at
that meeting, their current meeting would effectively be like the first
time. If this were true in general, even in encounters of many turns,
each turn would be like the one and only. Contests would have no
past and no future. This implies that cooperation is much more likely
to be sustained in small communities, with stable populations, in
which one is going to encounter the same people again and again. It

is here, rather than in large, anonymous cities, that one is likely to pay in the future for defecting in the present.

4. *For cooperation to emerge as the best strategy, there must be more than one player who is willing to cooperate.* To see this, imagine a collection of players all of whom defect every time. In this situation, TIT FOR TAT will cooperate on the first turn, thereby being exploited, and defect from then on. Indeed, it can be shown that no strategy involving cooperation can be successful on its own in a world full of defectors. If everyone always defects, the best anyone can do is defect also. This fact is very important. What it shows is that cooperation is not individually accessible. Applied to the sociobiological account of altruism, it means that a single mutant altruist in a population of selfish defectors will either be destroyed or be turned into a defector itself. For cooperation to dominate defection, a group of cooperators must appear together. Furthermore:

4a. *They must recognize each other.*

4b. *They must transact with one another in preference to transacting with unconditional defectors.*

Now the size of the needed cluster of cooperators does not have to be very large for cooperation to end up outdoing defection, especially if the gains from mutual cooperation are very large. But it is critical that cooperators be able to identify each other.

If all these conditions are met, cooperation can become the dominate mode of social interaction. If all these conditions are met, they seem to yield a functional explanation of cooperation. Cooperation can appear not as the product of rational, intelligent, *intention* to cooperate but as a result of pursuing individual interests in a world that is structured like a prisoner's dilemma game.

How likely it is that such conditions could be met? The participants in Axelrod's contest knew the rules of the prisoner's dilemma game, they knew they would be competing against others who knew the rules of the game, they knew that they would face these others on multiple occasions, they knew that at least most of these others would remember how they acted and respond to it, and they knew that since the point of the contest was to accumulate the most points, future interactions had to be weighed heavily in deciding what to do on any given turn. And even with all this knowledge, cooperative choices would only survive in a context in which there were other potential cooperators.

It is possible to imagine rational, intelligent actors in the real world developing some of this knowledge and thus arriving at a cooperative strategy. It is even easier to imagine actors who have genuine concern for the interests of others arriving at this strategy. And it is easier still to imagine actors living in a social context that values and fosters cooperation and reciprocity arriving at this strategy. But it is hard to imagine selfish genes, or fish, or birds, with no knowledge of strategy, no concern for the future, no concern for other genes, fish or birds, and no sensitivity to social norms, arriving at this strategy. Axelrod's work provides a reasonably convincing demonstration that cooperation can come to be dominant under the right conditions without any participants having it explicitly as a goal. Cooperation can be a stable strategy. But at the same time, it provides a convincing demonstration that cooperation is most unlikely ever to get started in the sociobiologist's world of selfish genes. It impels one to look for other factors that may be operating to promote cooperation in human social groups.

Furthermore, even if conditions favoring cooperation in the natural world may have existed from time to time, and some simple, self-seeking strategy like TIT FOR TAT may then have made cooperation a reality for one or another animal species, it obviously does not follow that cooperation in human social groups, when it occurs, is also the result of self-seeking TIT FOR TAT. For among the options that human beings have available to them that were precluded from the prisoner's dilemma tournament are these: (1) They can change the payoffs associated with various outcomes of the game. (2) They can change the game. (3) They can foster in individuals such high value for honesty and group solidarity that defecting would simply be out of the question, no matter what the potential individual benefit.

In essence, human beings, as members of cultures, are more like Axelrod than they are like the people who played in this tournament. As members of cultures, they have the power to set up the games, determine their rules, and specify their payoffs and penalties. Each of these capacities to specify the game is untouched by an analysis of how people act from within it. And it is this fact about people—that they are the products and makers of culture—that leads to the second major line of criticism of sociobiology as applied to people.

Culture and Biology

People, and so far as we know, only people, live in cultures. The institutions of culture and the existence of language introduce a source of influence of one generation on the next that is not genetic. Culture-induced influences on behavior progress much faster than genetic ones. Culture seems to dominate the genes.

To see the force of this claim, all we have to do is look around us. People live in a world of man-made objects. The environmental conditions to which they must and do adapt have not been around long enough for evolution to have produced the adaptations. And they won't stay around long enough for evolution to catch up. Furthermore, cockroaches the world over, if they are members of the same species, look and act basically the same. So do elephant seals, and whales, and robins. But people? All of a species, yes, but they look vastly different in different cultures. More to the point, they breed differently, raise children differently, fight differently and for different reasons, and form different kinds of social hierarchies. Human beings have no doubt evolved as a species and will continue to do so. But it would seem that evolution has to take a back seat in accounting for the variety of human social practices and their dynamics.

The sociobiologist can hardly deny that there is great diversity among human cultures. What he would suggest is that by looking beneath the surface—by looking deeply for the source of cultural institutions—people will see that they are all designed to meet the same, universal, biological "whisperings within": the whisperings that impel people toward the pursuit of inclusive reproductive fitness. People may, individually or culturally, deny it. They may deceive themselves into thinking their individual motives and social practices are governed by something other than the drive to perpetuate their genes. But if they look deeply enough, that is what they'll find—everywhere. Thus, sociobiologist R. D. Alexander says, "In terms of evolutionary history, human behavior tends to maximize the bearer's reproduction. Selection has probably worked against the understanding of such selfish motivations becoming a part of human consciousness, or perhaps being readily acceptable."

It is probably not possible to show that no matter how deeply people look, they will not be able to find inclusive fitness maximization at the root of some cultural practice. Probably, the sociobiologist will always be able to contend that people haven't looked deeply enough. What one can do, however, is make it clear that, at least sometimes, the goal of inclusive fitness, if present at all, must be buried very deeply indeed. The evidence that bears on this issue most tellingly comes from the anthropological study of patterns of kinship in different cultures.

The sociobiological understanding of kinship is clear. Kinship is biological/genetic. One's parents and siblings are half-kin, one's grandparents are one-fourth kin, one's cousins are one-eighth kin, and so on. Since the deep interest people have in kinship is supposedly biological, behavior toward kin should be influenced, at least implicitly, by degrees of genetic relatedness, and kinship patterns across cultures should respect this biological notion of kinship.

In the industrialized societies with which we are most familiar, they do. But in many other societies, they do not. An enormous panoply of kinship relations has been discovered by anthropologists, and some of them fly directly in the face of the biological definition of kinship. All cultures acknowledge kinship, and people in all cultures behave differently toward kin than toward nonkin, but the relation between kinship defined culturally and kinship defined biologically varies from culture to culture. And it makes no fitness-maximizing sense to be self-sacrificing to one's cultural kin unless they are also biological kin.

Some of the evidence on the cultural determination of kinship has been reviewed by anthropologist Marshall Sahlins, in his book *The Use and Abuse of Biology*. It is estimated that fully a third of the world's societies practice what is called "patrilocal residence." A man's son marries someone from outside the village, then resides with her as part of his father's family. When the son's children marry, the same pattern holds. And the same pattern holds for the son's brothers. Thus, a man's extended, collaborating family will include his brothers, sons, grandsons, cousins, and so on, along with their spouses. It will *exclude* his sisters and his daughters, although they are closer genetic relatives than his cousins, not to mention his in-laws. And in these societies, *close* relatives are defined by physical proximity, not genetic proximity.

Furthermore, what a man does for a close relative he will not do for a distant one, even if the "close" relative is a (male) second cousin and the "distant" relative is a sister. Now the sociobiologist might point out how a pattern of kinship like this serves an important biological function, by preventing incest, as indeed it does. He might further argue that this biological function is what explains it, that preventing incest really is in the interest of inclusive fitness. But there are surely other ways to prevent incest without devoting more resources to cousins than to daughters. One can pass laws against incest instead of casting daughters out. Thus, while preventing incest in this way may have adaptive value, it is surely not the way to maximize inclusive reproductive fitness.

In some cultures, a reversed pattern of kinship operates. In the Trobriand Islands, for example, kinship is matrilineal, and a child's biological father is not treated as family and contributes nothing to the child's support beyond its early years. Indeed, "fatherhood" is not interpreted biologically. Husbands whose wives become pregnant while they are away treat the coming baby as their child with equanimity. Anthropologist Bronislaw Malinowski offered the following account as evidence:

> One of my informants told me that after over a year's absence he returned to find a newly born child at home. He volunteered this statement as an illustration and final proof of the truth that sexual intercourse has nothing to do with conception.

Further examples of kinship practices that violate sociobiological expectations can be found in patterns of infanticide and adoption. Killing one's young does not, on the face of it, make any sociobiological sense. It is hard to see how such an action can serve inclusive fitness. Nevertheless, infanticide does occur in many different species, and among people, in many different societies. The sociobiological case that could be made is that by killing some of one's young the parent increases the likelihood that the others will survive, and survive with enough resources to be competitive in the reproductive game they will play as adults. What undoes this story, at least in the case of humans, is that there are several cultures in which infanticide is commonplace where adoption is *also* commonplace. Indeed, in these cultures there are countless families that have both killed at least one of their own offspring and adopted and raised as their own the children of others—often the children of slain enemies. What a practice like this does to serve inclusive fitness is hard to discern.

On the basis of evidence like this, it may well be that people are willing to make all kinds of sacrifices for their kin, but it is only in some cultures that people take kinship to mean anything like what biologists mean by it, and what sociobiologists *must* mean by it if their inclusive fitness-maximizing account of altruistic behavior is to be taken seriously.

There is a more general point to be derived from this discussion of kinship. Across cultures, kinship is not defined by some universal, physical principles. The *meaning* of kinship is something that is negotiated in each culture; its meaning is social, not biological. The same is true for other concepts that are of great importance to the sociobiologist. Indeed, in general, people are not moved by the physical characteristics of other people's actions, but by what these actions *mean*. And the meaning of an action is largely culturally determined. A threat in one culture may be a courtship gesture in another. An act of altruism in one culture may be an insult in another. The very categories "altruism," "threat," "insult," and so on may be construed very differently in different cultures. Sociobiologists assume that here is a tight, biological connection between actions on the one hand and the motives that underly them and meaning conveyed by them on the other. They believe that one can, as it were, read motives and meaning directly from action. Thus, they frequently suggest, social phenomena like war can be traced to biologically determined agressiveness and territoriality. But people don't fight wars out of aggressiveness, at least not always. People fight out of love, or honor, or guilt, or humaneness. As Sahlins puts it:

> It is . . . difficult . . . to conceive of any human disposition that cannot be satisfied by war, or more correctly, that cannot be socially mobilized for its prosecution. Compassion, hate, generosity, shame, prestige, emulation, fear, contempt, envy, greed . . . the energies that move men to fight are practically coterminus with the range of human motivations. And . . . the reasons people fight are not the reasons wars take place.

To forget this, to forget that the relation between motive and action is mediated by the meaning given to the action by the culture, is to make a serious error. Sociobiology elevates this error to the level of fundamental, theoretical principle. One must wonder what the sociobiological, fitness-maximizing account of sex, parenting, altruism, aggression, and other social activities has to offer when the meaning of such activities, and their particular manifestations, are not themselves biologically determined.

This is a very important point to keep in mind. Sociobiologists describe the behavior of animals with words like aggression, territoriality, entrepreneurship, slavery, jealousy, love, dominance, courtship, and caste—words that all have particular meanings in the context of human culture. The presumption is that they have the *same* meaning when applied to the behavior of animals. Indeed, it is almost irresistible to read about "courtship" in birds and invest that term with all that courtship means to people. The presumption is false, but it has a significant effect. By talking about "slavery" in ants, one transforms slavery from a social concept into a natural one. Slavery, like Mars, is out there in the world, ready to be observed; it is not a creation of human social institutions. To use words like these, independent of the social contexts that give them their meanings, is to give them the status of things, of culture-free, natural objects.

The argument we have been reviewing for the importance of culture is frequently misunderstood. Appeals to cultural determination of action are interpreted as meaning that when culture emerges, biology and things physical stop being relevant. Culture is caricatured as an impalpable, nonmaterial force that stops biological evolution in its tracks. Some cultural determinists may hold this view. They may believe that people evolved to a certain point and then culture took over—biology creating the conditions of its own obsolescence. But one need not subscribe to this view. Indeed, a far more persuasive case for the importance of culture has been made by anthropologist Clifford Geertz.

Geertz has suggested that much of the most recent and dramatic human biological evolutionary change—change characterized primarily by enormous growth of central nervous system size and complexity, including a tripling in size of the cerebral cortex—ocurred *after* people were living in cultures. What this means to Geertz is that the presence of culture, and people's dependence on it, were important environmental influences on the shape that the most recent human organic evolutionary changes took. As Geertz says, "Not only was cultural accumulation under way well before organic development ceased, but such accumulation very likely played an active role in shaping the final stages of that development." Thus, man without culture is not a savage, noble or ignoble. "He is literally a basket case, uprooted from the environment to which his biology is attuned." A fish out of water. Culture then is not something new, laid on top of biological evolution; rather, it is inextricably connected with that evolution. Geertz further suggests:

> Man is, in physical terms, an incomplete, an unfinished animal; what sets him off most graphically from non-man is less his sheer ability to learn (great though that is) than how much and what sorts of things he *has* to learn before he is able to function at all.

So the argument for the importance of culture need not be an argument that people are above biology. It need not imply that people have mastered the problems posed by the biological world. It need not be an example of anthropocentric hubris. People are dependent. They function under significant constraint. But the source of this constraint and dependence is less biology than culture. In this respect man is the feeblest of all animals, needing as he does a rich social structure to flourish—even to survive. Thus, to Dawkins' list of constraints on perfection must be added another; at least when talking about people. Human action is constrained by the social and cultural institutions in which people find themselves. If such institutions are less than optimal for fitness maximization, human behavior will be also.

It needn't always be the case that the rules or constraints that seem to govern cultures will be at odds with the rules sociobiologists have in mind when they talk about biologically determined, inclusive fitness maximization. The examples of kinship relations just discussed are striking because they are in conflict with the sociobiological story, but there are plenty of cases where culture and biology seem harmonious, and there is no principled reason why there could not be more. As a matter of fact, our own culture may be a prime example of the fit between culture and biology. We are living in a time in which the pursuit of self-interest in the free-market economy provides the primary metaphor for understanding social relations. As a result, our social and cultural categories overlap with our economic ones. And as we have already seen, our economic ones motivate our biological ones. Thus, there is at present a happy confluence of biology, economy, and culture.

This fit between our cultural rules and sociobiology may go a long way toward explaining why the sociobiological story seems so compelling. It does seem to capture salient features of modern life. *It does capture salient features of modern life.* But when patterns of cultural organization are given biological interpretation and explanation, a significant transformation occurs. What should be understood as the result of local, cultural, and historical contingency comes instead to be understood as universal, biological necessity. Cultural practices can be changed. People can analyze them, evaluate their consequences, and if they find them ineffective, inappropriate, immoral, or otherwise undesirable, they can alter them. Biological necessities cannot be changed. People are stuck with them whether they like them or not. This distinction echoes the discussion in Chapter 2 of what "it's human nature" means. If by "it's human nature" people mean natural, biological law, there's nothing to be done. If by "it's human nature" people mean the product of particular cultural characteristics, there's plenty to be done. It may be important to understand that, in the modern world, the pursuit of self-interest is the principal guide to conduct. But it is just as important to understand that the modern world is this way because *we,* not God or Darwin, have made it this way.

The Limits of
Behavior Theory

*Thus does the economy, as the dominant institutional locus,
produce not only objects for appropriate subjects, but subjects
for appropriate objects.*

MARSHALL SAHLINS

*O*ne of the criticisms of economics devel-
oped in Chapter 6 was that what economists regard as eternal characteristics of human nature are actually rather recent historical
developments, dependent on the existence of a market system that
encouraged people to pursue their individual, economic interests.
"Rational economic man" was an eighteenth-century invention, having little relation to how people actually behaved in societies prior to
that time. Once invented, the economic man concept led to the formation of social practices and institutions that were consistent with
it but that undermined previous forms of social organization. Thus,
the economic man conception was a self-fulfilling one. First, one claims
that deep down people are really such and such; then a world is created in which people can only survive if they act like they are such
and such; sure enough, in that world, people become such and such
deep down. Otherwise, they simply disappear.

If all we had to go on was economics, this little scenario might
have considerable plausibility. After all, human cultures do change,
and people's behavior can be influenced by how they conceive of

themselves. So modern economics really could be about modern, free-market man, and not universal man. Sociobiology raises some difficulties for this argument with its appeal to biological determination, but the relevance of sociobiology to human culture is, as we've seen, suspect or at least controversial. The key to undermining the historical argument lies in behavior theory. It bridges sociobiology and economics. It attempts to account for pigeons and rats in boxes and for people in schools and factories. Surely, it would be preposterous to suggest that the conceptions of behavior theory are also self-fulfilling. So while we can't perhaps be sure that economic man was not an invention, we can be sure that "economic pigeon" was not. And by inference, if pigeons really are economic, and if people allocate resources the same way pigeons do, then people are also economic.

A case can be made that "economic pigeon" is as much an invention of the behavior theorist as "economic man" was an invention of Adam Smith's. Indeed, it is the same invention. And indeed, the inventions of behavior theory actually contributed to the coming of age of the free market even before "behavior theory" proper existed.

The Limits of Scientific Theories

To begin to make the case, we must make a distinction between two seemingly similar but actually quite different claims:

Claim 1: Behavior is controlled by the principle of reinforcement.
Claim 1A: Behavior can be controlled by the principle of reinforcement.

The first of these claims is a statement of a universal, natural law. It says that people can understand *all* behavior, in all circumstances, as the product of the previous operation of contingencies of reward and punishment. (Actually, this is not quite true. Behavior theorists have long distinguished operant behavior from reflexes, a distinction that is roughly between voluntary and involuntary.) The second claim is much more restricted, or qualified. It says that there are certain circumstances in which reinforcement and punishment control behavior. Just how general this claim is depends on just how general these circumstances are.

Now the second claim is completely uncontroversial. One doesn't need behavior theory to know that (almost) every man has his price. If a person is hungry enough, he will do practically anything for food.

If a person is being tortured severely, he will do practically anything to have the torture stop. Less dramatically, if there is nothing else around to influence a person's behavior, he may well act so as to maximize reinforcement. This is hardly news.

Behavior theorists, however, argue that their research supports the first claim, not the second. Or, if the principle of reinforcement is not completely universal, then the circumstances under which it operates are much more widespread than people would have believed. The debates that occur between behavior theorists and their critics are not debates about the truth or falsity of the principle of reinforcement (for surely it is true) but about the generality of the circumstances of its applicability.

The same can be said about the debate between economists and some of their critics. Consider these two claims:

Claim 2: People act to maximize their economic self-interest.
Claim 2A: People can act to maximize their economic self-interest.

Again, the issue is not whether self-interest maximization is true. The debate is over whether the economist's evidence for self-interest maximization supports the unqualified claim (claim 2) or the qualified one (claim 2A).

It may seem to be nit-picking to make these distinctions, but if it were to turn out that claims 1A and 2A were actually the correct ones, rather than claims 1 and 2, some very important implications would follow. If behavior *can be* controlled by reinforcement, and if people *can be* self-interest maximizers, it seems natural to ask whether they *should* be. Should conditions in society be arranged so that claims 1A and 2A hold, or shouldn't they? Will such conditions make for a good society? Will such conditions make for good people? Is this the best way for people to live their lives? All of these questions demand moral judgments. They require people to have a conception of what a good life is so that they can decide whether the conditions that support claims 1A and 2A support that good life.

Now the principles of behavior theory and of economics can still be extremely useful under these circumstances. They can tell people what social conditions to establish (if they think a society in which claims 1A and 2A are true would be a good society), or what social conditions to avoid (if they think a society in which claims 1A and 2A are true would be a bad society). They can, in short, provide a set of techniques that will help people create a good society. But they decidedly cannot indicate what a good society should be. And this is a striking limitation in light of the efforts of economists and behavior theorists (and sociobiologists) to convince people that their scientific discoveries obviate these sorts of moral questions. They claim to be indicating how things inevitably are (claims 1 and 2), so that arguments about how they should be become rather idle. Economics and behavior theory are not meant to be moral arguments in defense of the free market; they are meant to replace such arguments.

Demonstrating that it is the qualified claims, and not the unqualified ones, that are supported by the available evidence will make a significant inroad into the major thrust of behavior theory and economics. It will provide critics of behavior theory and economics with considerable support.

But more is needed. For consider these two claims:

Claim 3: Objects fall toward the center of the earth when dropped.
Claim 3A: Objects can fall to the center of the earth when dropped.

Here is an example of the same type of dispute as before, only in physics. And in this case we know that claim 3A rather than claim 3 is true. Objects fall toward the center of the earth when dropped only under certain conditions. Gravitational force must be present. Rigid objects must be absent from the object's path when it falls. If the falling object is made of metal, there must be no strong magnets above it. There are plenty of conditions in which claim 3 would not hold. Indeed, consider these two claims, now from astronomy:

Claim 4: Planets revolve in elliptical orbits around the sun.
Claim 4A: Planets can revolve in elliptical orbits around the sun.

Even here, it is claim 4A and not claim 4 that is true. Only under certain circumstances (the ones that have obtained for millions of years) do planets revolve in elliptical orbits around the sun. One can easily imagine circumstances (and modern physics suggests that some of them will eventually obtain) in which our solar system would collapse, and claim 4 would be false. The point is that *all* scientific generalizations, no matter how universal and eternal they appear to be, are true only under limited conditions. All such generalizations are properly understood as being like claim 4A rather than claim 4. Often, when they state their generalizations, scientists don't bother stating what the limiting conditions are. They say something like,

"All other things being equal, planets revolve in elliptical orbits around the sun." In the case of something like planetary motion, we don't usually worry about what all those other, equal things are, because we can assume that for now, and for long into the future, other things really *are* equal. But this doesn't change the fact that even the most rock-solid scientific generalizations are true only under *some* circumstances. In this respect, there is nothing special about either the principle of reinforcement or the principle of self-interest maximization.

Does the fact that it is really only qualified claims that are true invite us to ask whether planets *should* revolve in elliptical orbits around the sun, or objects *should* fall toward the center of the earth when dropped? It surely wouldn't occur to anyone to do so, because in the modern conception of the nature of things it does not make sense to talk about inanimate objects in moral terms. Moral categories, for most of us, are restricted to people. But even if moral debate about planetary motion did not seem senseless, it would certainly seem pointless. Even if it is only under certain conditions that planets revolve around the sun, people are powerless to do anything about those conditions. The outcome of a moral debate about how planets should move would simply have no consequences. Thus, it is critical to the debate about behavior theory and economics that the circumstances under which people are reinforcement-governed, self-interest maximizing organisms are circumstances they can do something about, circumstances subject to human discretion, intervention, and alteration. Otherwise, the practical significance of the distinction between "is" and "can" would vanish. The behavior theorist and economist might be inclined to argue that, even though their principles are true under only limited conditions, they are the conditions that people presently live under and they came about through an evolutionary process because they were better at sustaining people than their predecessors. So current social conditions are the result of natural selection, not human discretion. To argue against this position it must be shown both that only the qualified claims are true and that the circumstances under which they are true are circumstances of our own making.

Limits of the Laboratory

Let us begin in the behavior theorist's laboratory, with pigeon or rat, pecking disk or pressing lever, for food or water. Does the confirmation of the principle of reinforcement obtained in this setting indicate what reinforcement does do or only what it can do?

Many years ago, two of B. F. Skinner's students, Keller and Marion Breland, decided to use their training in behavior theory to start a commercial venture. They used reinforcement principles to train animals to engage in a variety of entertaining stunts. People paid to watch the animals perform, or companies paid for "live" displays advertising their products or services. Their venture was extremely successful. They were able to produce well-controlled, amusing, and extraordinary displays of behavior from a wide range of different animal species, in a range of circumstances far greater in complexity than the researcher's experimental chambers.

Although over the years their successes far outnumbered their failures, there were failures. As the failures accumulated, the Brelands began to notice a pattern to them. In 1961, they published a paper, titled "The Misbehavior of Organisms." In that paper they recounted some of their failures and drew an important lesson from them. First, consider an example of a training failure. The Brelands were trying to train a raccoon to work for coins and then deposit those coins in a "piggy bank." What they got instead was a miserly raccoon:

> The response concerned the manipulation of money by the raccoon (who has "hands" rather similar to those of the primates). The contingency of reinforcement was picking up the coins and depositing them in a 5-inch metal box.
>
> Raccoons condition readily, have good appetites, and this one was quite tame and an eager subject. We anticipated no trouble. Conditioning him to pick up the first coin was simple. We started out by reinforcing him for picking up a single coin. Then the metal container was introduced, with the requirement that he drop the coin into the container. Here we ran into the first bit of difficulty: he seemed to have a great deal of trouble letting go of the coin. He would rub it up against the inside of the container, pull it back out, and clutch it firmly for several seconds. However, he would finally turn it loose and receive his food reinforcement. Then the final contingency: we put him on a ratio of 2, requiring that he pick up both coins and put them in the container.
>
> Now the raccoon really had problems (and so did we). Not only would he not let go of the coins, but he spent seconds, even minutes, rubbing them together (in a most miserly fashion), and dipping them into the container. He carried on the behavior to such an extent that the practical demonstration we had in mind—a display featuring a raccoon putting money in a piggy bank—simply was not feasible. The rubbing

behavior became worse and worse as time went on, in spite of nonreinforcement.

This raccoon was not simply failing to maximize reinforcement. It
was persisting in an activity that cost it food. And when the Brelands
tried to increase its incentive by making it hungrier, the misbehavior
only got worse. What could account for this failure of the principle
of reinforcement? The Brelands saw it, and other failures like it, as
an indication that powerful principles that overrode the principle of
reinforcement were operating:

> Here we have animals after having been conditioned to a specific learned
> response, gradually drifting into behaviors that are entirely different
> from those that were conditioned. Moreover, it can easily be seen that
> these particular behaviors to which the animals drift are clear-cut
> examples of instinctive behaviors having to do with the natural food-
> getting behaviors of the particular species.

Thus the raccoon manipulates the coins because the coins have
been associated with food, and the raccoon treats the coins in the
same way it treats food. It does this not through the operation of the
principle of reinforcement but in spite of it. The source of the behavior is instinctive. It is a built-in pattern of behavior triggered by
appropriate environmental stimuli, without regard to its consequences. This instinctive source of the animal's behavior represents
a candidate for behavior control to compete with the principle of reinforcement. It is a candidate sufficiently powerful that even when the
animal is hungry, and has already learned what to do to produce food,
and is in an environment very different from its natural one, to which
these instinctive behaviors are presumably attuned, it still dominates
control by the principle of reinforcement.

Many years after the Brelands reported their examples of misbehavior, Skinner acknowledged that sometimes the influence of reinforcement could be suppressed by what he called "phylogenetic"
(instinctive) influences. But, he argued, suppression could also go the
other way. Operant contingencies could suppress instincts. An example he cited involved a pigeon that had been trained to peck a disk at
a very high rate for food. The pigeon learned to peck very fast, but
then when the grain was made available, it continued pecking fast at
it—so fast, in fact, that it frequently failed actually to get any grain
inside its beak. Skinner further argued that in the case of people this
direction of suppression—of instinct by learning—was by far the more

common one. "Civilization has supplied an unlimited number of examples of the suppression of the phylogenetic repertoire of the human species by learned behavior."

Just so. But in making this point, Skinner was conceding a much larger one. For the real message of the Brelands' findings is not that the principle of reinforcement is false, or even that it isn't powerful. After all, even they used it successfully, and countless behavior theory researchers have used it successfully in their laboratories as well. No, the real message of the Brelands' findings is this: there is a very powerful influence on the behavior of animals other than the principle of reinforcement. This influence is so powerful that it will show itself under nonoptimal conditions, interfering with the food-getting behavior of very hungry animals. Any adequate theory of animal behavior will have to take account of this influence. Behavior theory does not.

Moreover, the behavior theorist's laboratory is constructed in such a way that influences other than the principle of reinforcement virtually never appear. Consider how distorted a view of the natural determinants of behavior one gets by looking for them in a setting that is designed to keep some very important ones from appearing. Thus, for the Brelands, the very success of the behavior theorist is a serious criticism of his experimental methods. He tests the claim that *all* behavior is governed by reinforcement in an environment from which the other serious candidate for control is excluded. Small wonder that his experiments confirm his claim. It is as if we set out to study human behavior in the house of the paralyzed man mentioned before, in the discussion of sociobiology. We might well conclude from such a study, understandably but wrongly, that legs had no significant function to serve for people—that they were vestigial.

The upshot of the Brelands' argument, then, is that the behavior theorist's research can only be taken as support for the claim that behavior can be controlled by the principle of reinforcement in circumstances in which instinctive sources of control are prevented from occurring. The behavior theorist's laboratory creates such a circumstance. It supports Skinner's argument that the principle of reinforcement can suppress other influences on behavior, at least under some circumstances. But at the same time, it invites us to ask whether it ought to.

The Brelands' findings are not an isolated anomaly. Over the years, many other phenomena like them have accumulated. What is the relevance of these findings for people? Are we to see ourselves, like raccoons, as collections of powerful instincts that modulate the influence of the principle of reinforcement? To do so would be to concede the case to the eager sociobiologist, who wants to make just such an argument. And how are we to explain the many successful applications of behavior theory principles in real-life human social institutions? Are these successes also the result of the suppression of instincts? No, to escape the Scylla of behavior theory without falling into the Charybdis of sociobiology, we must look for nonreinforcement influences on human behavior that are also noninstinctive.

Limits of Successful Applications

By now, successful applications of behavior theory principles are legion. Behavior theory principles have been used to treat severe problems of the sort that confine people to mental hospitals. They have been used to treat more mundane problems like bedwetting, bedtime tantrums, obesity, alcohol abuse, and smoking. They have been used in schools, both to eliminate unruly behavior and to facilitate learning. Indeed, some of the central features of computer-assisted instruction, which is becoming increasingly commonplace in schools, had their origin in behavior theory research. Finally, they have been used in a wide range of work settings, both industrial and white collar, to increase productivity.

Although we may not all be aware of it, the chances are pretty good that everyone has either experienced such an application or knows someone who has. Behavior theory principles have filtered down to influence everyday practices of childrearing, even though the people who are using these principles may be unaware of their origin—or even that they are using them. Don't all of these successful applications indicate that behavior theory really does capture something essential about people? Don't they inspire confidence that behavior theory principles will work wherever they are applied? And don't these successful applications constitute evidence to support the claim that behavior is controlled by the principle of reinforcement, rather than the claim that behavior can be controlled by the principle of reinforcement? If so, we can concede that behavior theory provides only an incomplete account of animal behavior without conceding its incompleteness as an account of human behavior.

These three questions have different answers. Yes, successful applications indicate that the principle of reinforcement says something essential about people. And yes, people should be confident that it will work in most situations in which it is applied. But no, this does not imply support for the unqualified claim that behavior is controlled by reinforcement. To see why this is so, we need to examine what successful applications imply about the basic principles that give rise to them, both in general and in the particular case of behavior theory.

Think about the various applications of the basic sciences of physiology and pharmacology that appear as modern medical treatment. First, consider the case of two widely used and effective drugs, insulin and aspirin. Insulin is used to treat diabetes. It is supplied to the body exogenously to make up for inadequate endogenous production. The physiologist would say that insulin supplied from outside performs just the same function, in just the same way, as endogenous insulin. Thus external insulin mimics the effect of internal insulin. And its success in controlling the symptoms of diabetes constitutes confirmation of the physiologist's understanding of the role played by insulin in normal body function.

Now consider aspirin. Aspirin has multiple effects. It alleviates pain, it reduces fever, it thins blood, it reduces inflammation, and so on. Now does aspirin work the way insulin does, mimicking endogenous bodily processes? Or does it work by dwarfing or overriding those processes? If there are endogenous processes that work like aspirin, it is not presently known what they are. That is, no single endogenous process that has all aspirin's effects—that aspirin can be said to mimic—has yet been identified. Thus aspirin's effectiveness in application confirms no particular theory of normal body function. Eventually it may, that is, eventually an endogenous process that aspirin mimics may be identified. But at the moment, aspirin is a technological achievement with no clear implications for the understanding of physiology. The fact that aspirin may work by overriding other things that are going on inside people in no way impugns its significance as an effective treatment. What it does impugn is the relevance of its effectiveness to physiological theory.

Now consider another example, artificial hearts. One doesn't need to know how the heart works to build an effective artificial heart. One needs to know the functions it performs, but not how it performs them. There is no reason to suppose that the most effective artificial heart will be one that mimics the natural heart in every way. All kinds of constraints and chance factors surely operated to influence how the natural heart evolved. It might turn out that the ideal blood-pumping machine has very little in common with the one that happens to inhabit the chest. Again, the mismatch between natural and artificial hearts has no bearing on the significance of an artificial heart as a technological achievement. All it shows is that such an achievement need not confirm any theory of physiology. Exactly the same sort of story can be told about dialysis machines. The good ones will perform the same functions that the kidneys perform, but not necessarily in the same way.

As a final example, removed from the domain of medicine, consider the digital computer. Computers perform feats of *artificial* intelligence. They store and retrieve data from memory, they perform various calculations, they draw pictures and transform them in various ways. They perform many of the functions that are thought of as attributes of human cognitive activity. Does the existence of the digital computer and its ability to perform various intelligent activities provide us with confirmation of a theory of human intelligence, or of human brain function? The answer is decidedly no. Computer circuitry is not like the human brain, and computers typically perform their functions in ways that are radically different from the ways that people perform the same functions. Computers work in ways that capitalize on features of their design (large memory, high speed of processing, and so on) that are importantly different (so far as is known) from features of the design of human cognitive capacity. So the development of computer intelligence need have no bearing on theories of human intelligence.

The point of each of these examples is the same: a successful application is not *necessarily* confirmation of any particular theory of endogenous processes. With regard to the successful applications of behavior theory, the question becomes; Do the applications work like aspirin or like insulin? Are they like artificial hearts or natural ones? Behavior theorists want to claim the latter. Applications are merely the systematic institution of the same contingencies of reward and punishment that operate on people haphazardly outside the applied setting, and always have. But unless the alternative possibility can be ruled out, there is no reason to take successful applications as support for the behavior theorist's unqualified claim. Instead, we could interpret applied successes in a fashion that exactly parallels the interpretation of laboratory successes. Applications work because the applied settings are constructed so as to prevent influences on behavior other than the principle of reinforcement from intruding. They work because the behavior theorist turns the mental hospital or the classroom into a human version of the rat or pigeon chamber. Under such circumstances, contingencies of reinforcement and punishment dominate or override other influences on behavior. In short, applications work with people under precisely the same circumstances in which they work with pigeons. They show that behavior can be controlled by reinforcement, as pain can be relieved by aspirin, or multiplication can be performed by a computer as a succession of high-speed additions.

The point here is not to prove that the principle of reinforcement *is* the behavioral equivalent of aspirin but merely to show that such an interpretation is not implausible. Such an interpretation does not impugn the effectiveness of the principle of reinforcement, any more than it impugns aspirin. What it does do is tell us that the behavior theorist will have to look elsewhere if he is to provide convincing evidence that the principle of reinforcement is unqualified in its applicability.

Where else can he look? We can take a lesson from physics. When the physicist derives general principles from research in the laboratory, he faces the same problem as the behavior theorist. Showing that Newton's laws of motion hold in a vacuum, or on a frictionless surface, is no guarantee that they have much to do with the motions of objects in the real world. Nor would some invention constructed in accordance with Newton's laws. Both the experiments and the inventions are human creations. What really convinces people that Newton's laws capture a significant piece of reality is that they see them exemplified in the behavior of natural objects, with no intervention from physicists. They are exemplified in the motions of the planets, and neither Newton nor his successors had much to do with that. So what one needs to find is some naturally occurring phenomena that exemplify the behavior theorist's principles without his intervention. Were we to find such phenomena, and were they widespread, we could confidently endorse the behavior theorist's unqualified claim.

And there are such phenomena to be found. All one has to do is look at the modern industrial workplace. Think about people working in, say, a clothing factory. Each of them performs a simple, repetitive task (for example, sewing on buttons, or cutting cloth), again and again, just as the rat presses the lever or the pigeon pecks at the disk. In addition, these tasks are essentially interchangeable in the sense that people do the work only for the wage it pays and would just as soon cut cloth as sew buttons if the wages were equivalent. They may be paid by the piece (what behavior theorists call *ratio reinforcement schedules*) or by the hour (roughly what behavior theorists call *interval reinforcement schedules*). And in the factory, behavior is extremely well controlled by the contingency of reinforcement. People do their work at high and predictable rates, just as the principle of reinforcement says they should. It is fair to say that the principle of reinforcement provides a good account of the behavior that occurs in the modern industrial workplace. Thus, at last, there is presumptive evidence that it is a good account of behavior in general.

So, the behavior theorist would argue, we really can account for behavior in the workplace and, indeed, for behavior generally. The factory is the behavior theorist's equivalent of Newton's planets. Or is it? How did the modern factory workplace come into being, and what preceded it? A look at the evolution of work reveals that factories aren't planets after all.

The Evolution of Work

The world into which Adam Smith injected his notion of economic man had very few factories. Recall how he extolled the virtues of the pin factory, in which division of labor and specialization of tasks greatly enhanced productivity. There would have been nothing worth noticing to extol if such factories had been a routine part of life. They were not. Factory organization of work is a relatively modern invention.

Centuries ago, what came to be modern industrial society was feudal. Large portions of land were controlled by lords. The majority of the population worked the lord's land, as serfs. These serfs had no legal alternative to the work they did. In return for his protection, serfs were required to work the lord's land and to turn over a fixed proportion of their yield to him. They had no choice of the terms they would work under or of the conditions of their work. They could not hire themselves out to the highest bidder. Nor could the lord sell off his land. The details of the relation between serf and lord were part of a longstanding set of political and social practices that was neither based strictly on economic considerations nor changed on the basis of these considerations. Neither lord nor serf acted so as to maximize marginal utility. Both land and labor were inalienable; that is, neither could simply be sold. This network of political and social practices is what Polanyi had in mind when he said, "Man's economy, as a rule, is submerged in his social relationships." These social relations exerted powerful constraints on the pursuit of self-interest.

If the factors operative in the choice of work were different in feudal than in modern times, so also was the nature of the work itself. Serfs, and other premodern workers, engaged in a wide variety of different activities in the course of a day. Their work required flexibility and decision making. The rhythm and pace of their work changed with the seasons. In addition, the work they did for the lord was integrated into the rest of their daily activities. They didn't leave home for the shop, work from nine to five, then return home to engage in personal pursuits. This pattern of work is in sharp contrast to the modern factory worker, who does the same thing all day, every day, with no flexibility or decision making required.

Over a period of many years, serfs were driven off the land in large numbers and eventually became wage laborers. This change coincided with other changes in work that resulted in the emergence of the factory system about which Adam Smith was so enthused. By the end of the eighteenth century in England, many of the descendants of serfs were not only working for wages but were free to hire themselves out to the highest bidder. Moreover, with increasing mechanization and division of labor, work became less and less varied and flexible. When the factory system was fully in place, behavior in the workplace seemed a perfect exemplification of the laws of behavior theory in operation.

The transition from feudalism to industrialization did not come all at once. For a time, even when masses of people were working for a wage, the wage they received, and the way they did the work, were determined by social custom, not by the competitive market. That is, workers did not hire themselves out to the highest bidder, and bosses did not try to extract maximal output for minimal cost. This transition has been documented by historian E. J. Hobsbawm, in his book *Labouring Men*. Gradually, says Hobsbawm: "Workers began to regard labor as a commodity to be sold in the historically peculiar conditions of a free capitalist economy; but where they had any choice in the matter, still fixed the basic asking price and quantity and quality of work by non-economic criteria." Finally, though, "Workers began to demand what the traffic would bear, and where they had any choice, to measure effort by payment. Employers discovered genuinely efficient ways of utilizing their workers' labor time."

Thus, complete control of work by wage rates (reinforcement rates) was not characteristic of early industrialization. This is not to suggest that work was uninfluenced by the reinforcement contingency. Clearly, if workers received no pay at all, they would not have worked. However, pay rates did not exert the same kind of control over workers as reinforcement rates exert over animals.

But wait. Hobsbawm said that workers and bosses both had to *learn* the "rules of the game." Perhaps this period of transition was transitional just because learning took some time, The pigeon has to learn what to peck and how often to peck it. The experimenter has to learn what kind of contingency to establish to get high response rates. So it may have been with bosses and workers. The emerence of utility-maximizing behavior on the part of both simply took some learning time, some time to adjust to new contingencies.

Indeed, one might want to argue, as the behavior theorist and economist surely would, that people were always maximizing utility, even when they were serfs. It is just that under feudal social organization, the reinforcers were different. They included protection, security, a place in a community, and so on. The change from feudalism to capitalism was a major social upheaval, a major alteration of environmental conditions. As biologists know, it takes time for an organism's behavior to evolve to meet new demands imposed by a changed environment. So perhaps the change in work that evolved slowly with the change from feudalism to capitalism was the evolution of a new, *specific* solution to the continuing problem of utility maximization. How else can we explain this transition?

Well, another way to understand the transition is this: Prior to industrialization, work was largely governed by social custom. One worked in such and such a way, not because that was the way to maximize, but because that was simply the customary way of performing the work. Then, early in industrialization, work was governed by *both* wages and social custom. The work one sought was governed by the available wages, but the manner in which one performed the task was dictated by the way that the task had been performed by previous generations. Finally, the last stage of the transition was one in which wages came completely to dominate both the work one chose and the way one did it. And the reason that work came to be completely dominated by the wage is that its principal competitor for control—social custom—had been systematically and intentionally eliminated.

If this point of view could be sustained, then the emergence of the factory could be seen in a way that parallels our understanding of the successes of behavior theory in the laboratory. The principle of reinforcement completely controls the behavior of animals in the laboratory only when, and because, other potential influences on their behavior—biological ones—have been eliminated. The principle of reinforcement completely controls behavior in the factory only when, and because, other potential influences on behavior—sociocultural ones—have been eliminated. If this can be shown, then the factory becomes just another successful application of behavior theory principles, telling no more about the general validity of behavior theory than a weight control program does. The factory becomes a case not like the planets but like airplanes, artificial hearts, and computers.

And it can be shown. For a central component of the final stages of development of the workplace, in its modern form, was a movement explicitly designed to eliminate custom as an influence on behavior. The movement was one of the earliest examples of what is now called "human engineering." It went by the name of "scientific management," and its founder and leader was Frederick Winslow Taylor.

Taylor argued that custom interfered with efficiency and productivity. What industry needed was a set of techniques for controlling the behavior of the worker that was as effective as the techniques used for controlling the operation of machines. Accomplishing this control involved two distinct lines of human engineering. First, one would need to discover the rates and schedules of pay that resulted in maximal output. Second, one would need to break up customary ways of doing work and substitute for them minutely specialized and routinized tasks that could be accomplished mechanically and automatically. The idea was to strip work down to its simplest possible elements, to eliminate the need for judgment and intelligence, and to wrench work free of its customary past. With this done, there would be no possible source of influence on work except for the schedule of pay. And the schedule of pay was something the boss could control.

Taylor made these objectives explicit in his book, *Principles of Scientific Management*. Published in 1911, this book summarized Taylor's approach to human engineering. In it he said:

> Owing to the fact that the workmen in all of our trades have been taught the details of their work by observation of those immediately around them, there are many different ways in common use for doing the same thing, perhaps 40, 50 or a hundred ways of doing each act in the trade. Now among the various methods . . . used in each element of each trade there is always one best method . . . which is quicker and better than any of the rest. And this one best method . . . can only be discovered through a scientific study and analysis of all the methods

. . . in use, together with minute, motion and time study. This involves the gradual substitution of science for rule of thumb throughout the mechanic arts.

With workplace tasks "scientifically" divided in this way, attention could turn to the manipulation of pay schedules. And they were manipulated. Piece rates, day rates and various combinations of them were tried. When the dust cleared, the ultimate method of compensation, according to L. M. Gilbreth, one of Taylor's disciples, was called "differential rate piece work":

> This consists, briefly, in paying a higher rate per piece . . . if the work is done in the shortest possible time and without imperfection, than is paid if the work takes a longer time or is imperfectly done. . . . This system is founded upon knowledge that for a large reward, men will do a large amount of work. The small compensation for a small amount of work may lead men to exert themselves to accomplish more work.

As one reads descriptions of the various pay schedules explored, in a book written by Gilbreth in 1914 (*The Psychology of Management*), one is struck by the parallels between it and another book, published by Skinner in collaboration with C. B. Ferster in 1957. That book, called *Schedules of Reinforcement*, is one of the signal achievements of behavior theory. It is a detailed demonstration of the lawful effects of different patterns of reinforcement delivery on the pecking of pigeons. Thus scientific management constitutes an application of behavior theory principles to the workplace—fifty years before behavior theory had discovered them!

Scientific management in general, and Taylor's version of it in particular, gradually faded from the industrial scene, partly because of worker resistance to its application. However, it has had a lasting influence on the structure of the workplace. According to Harry Braverman, in his book *Labor and Monopoly Capital*, "It is impossible to overestimate the importance of the scientific management movement in the shaping of the modern corporation and indeed all institutions which carry on labor processes." This view is echoed by management consultant Peter Drucker, who says, "Scientific Management is all but a systematic philosophy of workers and work. Altogether, it may well be the most powerful as well as the most lasting contribution America has made to Western thought since the Federalist Papers."

Thus, behavior theory provides a good account of behavior in the workplace. It does this not because work is a natural exemplification of behavior theory principles but because behavior theory principles, in the form of scientific management, had a significant hand in transforming work into an exemplification of behavior theory principles. Polanyi's judgment on Adam Smith could apply as well to behavior theory: "No misreading of the past ever proved more prophetic of the future."

The demonstration that the modern workplace for which behavior theory is so well suited is an historical peculiarity that behavior theory helped create does not leave the behavior theorist without response. The response would appeal to cultural evolution. Skinner has suggested that the principle of reinforcement operates at the level of culture as well as at the level of individuals. The cultures that develop practices and modes of organization that best meet the contingencies imposed on them by the environment are the ones that survive and flourish. The reorganization of the workplace, in keeping with the principles of behavior theory, is merely a sign of cultural selection by the principle of reinforcement. Modern, industrial society is no cultural accident. It is the most efficient and productive system ever known. It was, after all, this amazing productivity that impressed Adam Smith about the pin factory. And it was the name of productivity and efficiency that Taylor set about to reorganize the workplace. The historical circumstances that led to this direction of cultural development may have been accidental, but the factors that nourished and sustained it were not. The industrial revolution is still with us because nothing can match it.

The thrust of this line of argument is to make the historical speculation of the last several pages appear somewhat idle. Ultimately, by natural law, only the good (efficient, effective, adaptive) social practices survive. The principle of reinforcement sees to that. This, recall, was Darwin's line on moral practices. Natural selection sees to it that only cultures possessing a good (adaptive) moral sense survive. And what is important on this line of analysis is not how a cultural practice begins—not the nature of the social mutation that gives rise to it—but whether it lasts. Scientific management may be seen as the elimination of superstitious, inefficient, misguided ways of working and their replacement by scientific, efficient, and productive ones. Culture simply evolved in an adaptive direction, guided by the principle of reinforcement.

In our critical discussion of sociobiology, we discovered that the evolutionary explanatory framework encourages a kind of Panglossian picture of evolutionary change. Given some bit of behavior, it is always possible to spin a tale about how that behavior is adaptive—indeed, about how it is the most adaptive thing an organism could do. If the limitations on such tales are just the limitations of the tale teller's imagination, the evolutionary framework becomes rather empty. One can account for the evolution of anything just by telling the appropriate evolutionary tale. But it is possible to give evolutionary accounts real teeth by taking seriously a set of constraints that operate on organisms as they evolve. One of those constraints is that evolution is *history,* that accounting for the evolution of a given trait observed in the present requires not only showing that it has adaptive value in the present but also spelling out the history of its evolution. That is, the evolutionary biologist must show us how it was possible "to get here from there." This frequently involves showing how intermediate forms in a trait's evolutionary history had adaptive value themselves.

We can demand the same of the behavior theorist and economist. It is not good enough to argue that modern industrial society evolved because it was "the best of all possible worlds," even if it is agreed that it is the best of all actual current candidates. No, to take this argument seriously, we also want to be shown that "intermediate forms" on the path from feudalism to modern industrialization were themselves adaptive, that they constituted more efficient and productive alternatives to what they replaced.

Economist Stephen Marglin, in a paper titled "What Do Bosses Do?" has attempted to show that the case *cannot* be made that intermediate forms of industrial organization evolved on the basis of efficiency and productivity. He has argued that key developments in early industrialization had almost nothing to do with productivity and efficiency. Instead, what industrialization and division of labor did was to give the boss much greater control of the production process than he had previously had. And the point of this control over production was not to increase productivity but to increase profit. Marglin's arguments are sufficiently provocative and significant to warrant presentation in some detail.

Consider Adam Smith's pin factory. Surely, there was an efficiency advantage in breaking pin production down into component tasks and in performing each of those tasks hundreds or even thousands of times before moving on the next one. Each job has a certain "set-up" time, and the longer one stays at the job, the more the costs of setting up recede to the background. Also, the worker can get into a rhythm so that he can do the job with less and less fumbling around as he continues. These are some of the advantages to which Smith referred in exalting the division of labor:

> This great increase of the quantity of work . . . is owing to three different circumstances; first, to the increase of dexterity in every particular workman; secondly, to the saving of the time which is commonly lost in passing from one species of work to another; and lastly, to the invention of a great number of machines which facilitate labor and abridge labour, and enable one man to do the work of many.

So breaking work up into subtasks is certainly efficient. But there is no appreciable gain in efficiency by having each of these component tasks done by *different* people. Timesaving, skill development, and machine invention would all be possible even if a single person performed each of the needed subtasks. Think about some everyday tasks that people perform. In preparing a stew, it is likely that cooks first peel and cut all the needed potatoes, then peel and chop all the needed onions, then peel and slice all the needed carrots. They don't do an onion, a potato, a carrot, an onion, a potato, a carrot, and so on. But there would be no gain in overall efficiency if different people cut each vegetable. And the invention of the food processor did not depend on the division of cooking among people. Or think about painting a room. Most people do all the sanding, scraping, and patching, then paint all the woodwork, then paint the walls. One could scrape and patch a little, paint a little, scrape and patch a little, paint a little, but that would be inefficient. But again, there is no overall gain in efficiency if different people do each subtask, nor did the invention of the electric sander depend on this sort of division of labor. The cases could be multiplied endlessly. It may be that some tasks require so much skill that it is unreasonable to think that a single individual could become a master of all the components. In such cases, division of labor among people makes sense. But many, many tasks, especially those performed in a "scientifically managed" factory, do not require that degree of skill. In these cases, we cannot explain the social division of labor by appealing to efficiency.

Indeed, if anything, dividing a task among different people *decreases* efficiency, since the efforts of different people need to be coordinated by someone who could otherwise be directly involved in production. It is just this need for coordination that gives the boss the crucial and controlling hand in the production process. It is in his hands, and his hands alone, that the finished pin resides. And that fact, according to Marglin, was the real motivation for industrial development as we know it.

Furthermore, Smith's idea that somehow invention is facilitated by the social division of labor is surely wrong. It is the person with a view of the task as a whole who is most likely to have imaginative ideas about what a good invention will be. Imagine being faced with the task of designing an elephant mover. Does one consult the dozen blind men who happen to be surrounding the elephant, each manually exploring a different part, for advice about the best design? Of course not, for none of them has an accurate conception of what an elephant actually is. And what happens to inventiveness when people are faced with a breakdown in the production process? What can the onion chopper do if there aren't enough onions? Or the potato cutter? It is only the "stew maker" who can improvise, by substituting turnips, or peppers, or what have you. But who is the "stew maker" when cooking has been divided into many hands?

All of this seems obvious. The more anyone knows about all aspects of an operation, the more likely he is to keep it afloat when things go wrong, or to design significant improvements in it. Yet few modern American factories are organized to take advantage of this "obvious" fact. And when people observe modern, foreign industrial operations, which are increasingly organized to involve individual workers in larger and larger chunks of the production process, they marvel at the productivity and reliability of their methods and talk about bringing this new "industrial revolution" home to America.

In buttressing his argument that control, not efficiency was behind the development of the factory, Marglin marshalls other evidence. First, in industries like the coal industry, owning the mine was sufficient to ensure control of production. A worker simply could not set up a competing coal-mining operation down the street. And in the coal industry, specialized division of labor did not occur. Second, against the argument that it was only in large-scale factories that one could afford to invest the capital necessary for building or purchasing expensive machinery, Marglin shows that the development of expensive machinery did not occur until *after* the factory system was firmly in place. Early on, workers used the very same methods in the factory that they could have used at home. The key differences between the two were that workers only participated in a small part of the production process in the factory, while they did it all at home, and that they worked to the bosses' schedules in the factory while they worked to their own at home. Indeed, in the wool trade, home-based methods of production were in use in the factory for over one hundred years after the factory was in place.

Once factory organization was developed, technological change was guided by its presence. It is indeed impossible to look at the modern factory and imagine the production processes it employs being used by independent individuals at home. But the point here is that the growth of technology was shaped by the factory, and not the reverse. It is true that the large-scale factory is what permitted the boss to accumulate enough capital for investment in expensive machinery. But it has never been shown that big, expensive machinery is the route to maximally efficient production. For all we know, industrialization *could have* evolved on a completely different but equally efficient path involving small-scale, worker-controlled production processes.

This really is hard to imagine with the smokestack skyline so much a part of everyone's life. To help see the point, consider cooking, an activity that is quite the opposite of factory production. Cooking is done on a small scale, by each family, on its own. Suppose that four hundred years ago, some feudal lord had decided that he wanted all of his serfs eating together. He might have constructed a massive dining hall, with a huge kitchen. When meals were prepared, different people would have performed different subtasks—carrot peeling, onion chopping, and so on. There was no technology, but there was social division of labor. Now, as technology evolved, it would evolve to meet the existing pattern of meal production. Everything would be huge. Food would be sold in enormous packages. No one would know enough about all aspects of meal preparation to set up an independent kitchen. With a technology like this in place, large-group meal taking would filter down through the ages, even as feudalism faded away. Independent, twentieth-century families, living in their own houses, would gather for communal meals. It would be hard for anyone to figure out an alternative. Is this mode of meal taking more efficient than the one we actually know? Perhaps yes, and perhaps no. But the point is that it certainly would be more efficient if technology had evolved to fit it. That is the point that Marglin is making about factory organization.

Perhaps the clearest evidence that efficiency and productivity were not the central justification for the factory system was the enormous difficulty that bosses had in harnessing the efforts of their workers. Workers chafed at the confining discipline of the factory. They malingered, they failed to appear, they quit altogether. Harnessing the worker was difficult, and for the successful boss it was a singular achievement. This is evident in the words of Andrew Ure, writing in 1835 in admiration of Richard Arkwright, one of the successful pioneers of factory organization:

> To advise and administer a successful code of factory discipline, suited to the necessities of factory diligence, was the Herculean enterprise, the noble achievement of Arkwright. Even at the present day . . . it is found nearly impossible to convert persons past the age of puberty . . . into useful factory hands. After struggling for a while to conquer their listless or restive habits, they either renounce the employment spontaneously, or are dismissed by the overlookers on account of inattention.

Why, if the factory was so unpleasant, did workers enter it to begin with? They were under no compulsion to do so after all. The economist would see this as a free-market choice. Workers vote with their feet, and if they marched into the factory, it must have been better than the alternatives. Well, Marglin points out that most early factory workers really had no effective choice. They were country people who had been driven off their land, or paupers, or disbanded soldiers, or women and children who had been offered to the factory by their husbands and fathers. Eventually, the problem of inducing workers to put up with the conditions of the factory disappeared. For as Marglin notes:

> Recruiting the first generation of factory workers was the key problem. For this generation's progeny the factory was part of the natural order, perhaps the only natural order. Once grown to maturity, fortified by the discipline of church and school, the next generation could be recruited to the factory with probably no greater difficulty than the sons of colliers are recruited to the mines or the sons of career soldiers to the army.

So it is that what for one generation is the wrenching out of a complex network of customs and social relations is for the next "only natural." So it is that scientific managers could see themselves as merely increasing the efficiency of work rather than transforming its very character. And so it may be that the behavior theorist, looking around at the "natural" order of things, can see his principles as reflecting an eternal, value-neutral necessity of human nature rather than an historical, value-laden contingency.

The notion of value becomes prominent if we take seriously the idea that the factory system is justified in terms of efficiency. For the economist, "efficiency" has a rather precise meaning. One system is more efficient than another if it produces greater output with the same inputs. Inputs include raw materials, capital, and labor. Now it is perhaps not very difficult to evaluate the amount of capital and raw materials that two production systems require. But what about labor? How is it evaluated? The simple way is just to look at the wage bill: the cost of labor is what one pays the laborer. But what about the cost of labor *to* the laborer? What is the cost to an individual of subjecting himself to year after year of mindless work and strict factory discipline? What is the cost to society of having millions of its people spend their productive years in this way? If these costs are treated as part of the input to production, production systems that otherwise seem quite efficient might turn out to be disastrously inefficient. But in order to include these costs as inputs—in order to have a richer understanding of labor costs than merely the wage bill—we must have some ideas about what sorts of activities make human lives valuable. The economist, or behavior theorist, in arguing for the efficiency of the factory, is arguing from an implicit theory of about the value of human activity. The value of human activity is simply the price one has to pay to hire it.

Turning Thinking Into Factory Work

I have tried to make the case that behavior theory principles do not so much find order in chaos as create it. What one knows from looking at the laboratory and at the workplace is that behavior *can,* under a limited set of impoverished circumstances, be completely controlled by reinforcement, not that it ordinarily or typically is. I have also tried to suggest that the distinction between "can" and "is" is an important one. For once it is understood that behavior theory is only indicating what *can* be the case; people must decide whether it *should* be the case. Moving into the domain of "shoulds" raises considerations that behavior theorists (and economists and sociobiologists) have tried desperately to avoid. Deciding on what should be depends upon having a moral theory, as a guide in determining how people should live. And these sciences, by their own admission, have nothing to say on this score.

Was the argument successful? It depended upon a particular interpretation of the history of the workplace—on the influences that helped give work its modern form. It tried to show how and why people attempted to create working conditions in which work would be completely governed by the rate and schedule of wages. It tried to suggest that these people succeeded and, in so doing, changed the nature of work. What is needed to make the historical analysis plausible is a laboratory demonstration that corresponds to it.

There are such demonstrations. One set of them can be summarized as "turning play into work." People are given the opportunity to engage in a variety of activities that might be regarded as pleasurable: solving various puzzles, for example. These are activities people would happily engage in in the absence of any reinforcement. Indeed, the activities are themselves reinforcing, so that people would do other operant things for the opportunity to participate in these activities. The twist in these demonstrations is that even though no reinforcement is necessary to keep people at the activities, they get it anyway, typically in the form of money.

And the reinforcement has two effects. First, predictably, it gains control of the activity, increasing its frequency. This might be called the "Boy, I would have done this for nothing, it's so much fun. But they're paying me on top of it. I'm just raring to go" effect. Second, when reinforcement is later withdrawn, people engage in the activity even less than they did before reinforcement was introduced. The withdrawal of reinforcement doesn't simply reduce responding to its prereinforcement, baseline levels; it eliminates responding almost completely. This might be called the "I used to do this because it was fun; then they started paying me, and now it just seems like work" effect.

Many demonstrations like this are collected in a book edited by M. R. Lepper and D. Greene called *The Hidden Costs of Reward.* In one demonstration, the experimental subjects were nursery school children. They were given the opportunity to draw with felt-tipped drawing pens, an activity that seems to have almost unlimited intrinsic appeal to young children. After a period of observation, in which experimenters measured the amount of time the children spent playing with the pens, the children were taken into a separate room where they were asked to draw pictures with the pens. Some of the children were told they would receive "good player" awards (reinforcement) if they did the drawing; others were not. A week later, back in the regular nursery school setting, the drawing pens were again made available, with no promise of reward. The children who had received awards previously were *less* likely than the others to draw with the pens at all. If they did draw, they spent less time at it than other children and drew pictures that were judged to be less complex, interesting, and creative. Without the prospect of further awards, their interest in drawing was only perfunctory.

What is important to note about this demonstration is that it is not an example of the failure of the principle of reinforcement. On the contrary, the awards seemed to gain control over the behavior. If they had continued to be available for drawing in the classroom, there is little doubt that high rates of drawing would have been maintained. The point of the demonstration is that prior to the introduction of the reinforcement contingency, something else was influencing the drawing, and that other influence was suppressed or superseded by the reinforcement contingency. This is a case of reinforcement working like aspirin, not like insulin. In similar fashion, the introduction of reinforcement contingencies in the factory may have contributed to the suppression or superseding of other influences on work.

Another series of experiments indicating that reinforcement could usurp control of an activity from other sources was conducted in my laboratory. Imagine someone seated in front of a matrix of light bulbs, five across by five down. Beside her are two push-buttons and a counter that keeps track of her score. Periodically, the top left bulb in the matrix lights up, signaling the start of a trial. "This is a game," she is told. "By pushing the two buttons, you can change the position of the illuminated light in the matrix of lights. If you do it right, you get a point. What I want you to do is to figure out the rules of the game; figure out what you have to do to earn a point."

She pushes the left button, and the light moves down one position. She pushes the right button, and the light moves across one position. Left, right, left, right. Down, across, down, across, one more left and one more right and the bottom right light is now lit. All the lights go out, and she gets a point. The next trial begins, and she pushes the buttons in exactly the same order and again gains a point. She does the same thing, with the same result, on the next trial. "Aha" she says, "this is easy. I know the rules of this little game." She calls over the experimenter and tells him: "You have to start on the left and alternate between left and right. Four alternations get you a point." "Wrong" says the experimenter, "try again."

She realizes she has made a silly mistake. The experimenter wants to know what one has to do to get a point—what is necessary. All that she has discovered is one particular way to do it. But there is no reason to believe that it is the only way. It is good enough—*sufficient*, to alternate between left and right, but it may not be *necessary*. How, then, does one go about finding out what is necessary?

The way to find out what is necessary is to vary what one does on each trial, in systematic fashion. To test whether it is necessary to alternate, starting on the left, one starts a trial on the right. If one's guess about the rules of the game is correct, this won't earn a point. But it does. So starting on the left is not necessary. Perhaps any alternation will do. If that's right, then a pattern of two left, two right, two left, two right won't work. Or maybe it's any regular pattern. If so, three left, two right, one left, and two more right won't earn a point. Maybe any pattern earns a point. This can be tested by just pushing away on one of the buttons.

The name of this game is "experimental science." There is a phenomenon, the getting of points, and the task is to explain it, to discover its causes. One goes about this task by formulating guesses or hypotheses and by doing experiments to test the hypotheses. The process one goes through in attempting to discover the rules of the game is precisely the one that scientists go through as they attempt to discover the rules of whatever "natural game" they are studying. And the process has two essential ingredients: formulation of hypotheses and tests of the hypotheses by systematic variation in experiment.

Moreover, there is nothing about these processes that is unique to science. People engage in them frequently, if somewhat less systematically, in everyday life. Someone interested in learning how to bake bread might find a recipe, and follow it carefully, step by step, with the bread turning out delicious. Now if all one cares about is knowing *a* way to make good bread, he will follow this recipe every time the need for bread arises. Similarly, if all our game player cares about is finding *a* way to get points, once she finds that left-right alternation succeeds, she will never deviate from it. But suppose one wants to know more than *a* way to make good bread. Suppose one wants to make different kinds of bread. Or suppose one wants to discover the most efficient way to make good bread. Now, what the baker must know is the essentials of good bread baking, what is necessary to make good bread—the rules, as it were, of bread baking. And as in the case of the button-pushing game, this requires experimentation, the formulating and testing of hypotheses. Steps in the recipe are omitted, combined, reordered; different kinds of flour or shortening are substituted. Eventually, in this way, one can discover which steps are essential, which are nice but inessential, and which are completely unimportant.

Now back to the little game. College students were recruited and told to try to discover the rules of the game. When they succeeded in discovering a rule, it was changed, and the students did it again. They were given no instructions about how one might most effectively tackle the problem. Sometimes, the students were told that they would get a few cents for each point they earned in the process of discovering the rule. Sometimes, they were told they would get a dollar for each rule they discovered. Sometimes, they were able to get a few cents for every point and a dollar for every rule. Finally, sometimes no monetary rewards were available at all.

Think about how these various contingencies of reward might affect a student's behavior. Suppose he could earn money for every point. This would put him in conflict. If what he cares about is discovering the rule, he will vary his responses from trial to trial. But if what he cares about is earning as much money as possible, once he finds a successful sequence of responses, he'll stick with it. After all, every time he experiments by varying his sequences of responses, he risks failing to earn a point. When one experiments with bread-baking technique, one risks producing lousy bread. If, in contrast, he only earns money by discovering the rule, then there is no conflict. Both the contigency of reward and the intrinsic demand of the task itself encourage him to vary his response sequences systematically and intelligently.

When students played the game, these varying conditions of reward made no difference at all. In all cases, students varied their responses from trial to trial with great efficiency. They used information from hypotheses that had failed to shape the new ones they would test. They were not perfect, but they were close. Almost every one of them discovered each of the rules, and they did so quite rapidly. The reinforcement contingencies failed to have any impact.

There is nothing terribly surprising in this. These students attend a highly selective institution. They are intelligent, and they have achieved great success in their studies before. Solving problems like these is something they are experienced at and good at and enjoy doing. They almost certainly would not have been admitted to the college they attend otherwise. Thus, perhaps it is not that reinforcement failed to control their behavior but that the reinforcement that resulted from discovering the rules was so potent that it overwhelmed the small sums of money that were available. What is reinforcing, remember, is an individual affair. With different people the money might have been more significant than the puzzle solving.

But there is more to the story. Another group of students was exposed to the same set of problems with the same contingencies of reward as the first set. What distinguished the two groups was that this second group had had prior experience playing the game. The prior experience was this: they were brought into the laboratory, shown the game, and told that every point they scored would earn them two cents. They were then given one thousand opportunities to play the game (trials). Note that there was nothing in the instructions they received urging them to discover the rule. They could have tried to if they wished, of course, but they could also just find a sequence of responses that worked and stick with it, earning as much money as possible.

And that is what they did. Perhaps not even realizing they had a choice, each student settled on a particular sequence of responses that occurred on almost 90 percent of all trials. Once one finds a way to make good bread, why fool around? Some of them seemed only half-awake, mechanically stamping out sequences while they daydreamed, or thought about their studies, or about the state of contemporary theater, or who knows what. The little game and the contingency of reinforcement had turned them into assembly-line workers, engaged in the same task, done the same way, over and over again, completely controlled by the reinforcement contingency.

What happened then when these newly formed factory workers were instructed to discover the rules? Did they go back to being smart and inquisitive college students? No they did not. Compared to the first group, they were much less effective at discovering rules. They discovered fewer of them and took longer in discovering the rules when they were successful. They seemed to test hypotheses more or less at random, failing to use past mistakes to inform and direct future efforts. And unlike the first group, what they did was powerfully influenced by the prevailing contingency of monetary reward. They were especially ineffective at discovering rules if each point they got earned them money. While they did not persist at the stereotyped sequence of responses that they had been using previously, neither did they completely abandon the mechanical orientation to the task that had served them earlier. They behaved like half-hearted scientists, people who more or less knew how to go about finding something out, but didn't care enough about the answer to make a maximal effort.

That we obtained this result is surprising, for several reasons. First, we were dealing with a group of puzzle solvers, with people who get their kicks from acting like scientists. Nevertheless, pretraining seemed to overcome this orientation, or at least to place the game outside of the problem-solving domain. (Even bright college students, after all, don't treat everything as a puzzle to be solved. They don't seek the general principles of shoe tying or tooth brushing.) And it did so even though the sums of money up for grabs were rather trivial.

Second, we might have expected their previous experience to make them better rather than worse at discovering the rules. No doubt they had tested and rejected some hypotheses while developing their stereotyped response sequence. For example, they probably learned that how fast they pushed the buttons, or which hand they used, made no difference. The inexperienced subjects didn't know these things when the rule discovery task began. In a similar fashion, someone who sets about to discover the essentials of bread baking *after* having had experience following a few recipes has a leg up on someone who is completely inexperienced, even if it never occurred to the experienced person to experiment while following the recipes. Finally, one of the rules that the pretrained students had to discover was the very same rule that had been in effect during their one thousand trials of pretraining. But even here, they were less effective than students who came to this problem completely naive.

Thus we have an experimental simulation of the historical process I argued for earlier. Exposure to contingencies of reinforcement creates an efficient, stereotyped pattern of behavior that can be executed with effortless and mechanical precision, just like work on the assembly line. The contingency makes it possible for people to do the right thing, over and over again. Lapses of attention have no cost, because attention is not required. Lack of intelligence has no cost, because intelligence is not required. The people become an extension of the machinery, a realization of Taylor's vision. Behavior can be completely controlled by the contingencies of reinforcement.

That this automatization is achieved at the expense of another potential influence on the nature of the activity becomes apparent when these students are later asked to discover rules. Students without pretraining know what rule discovery means. They have been participating in a tradition of rule discovery or problem solving for years, and it is a relatively simple matter to plug this new challenge into one's traditional wisdom from previous ones. The pretrained students are a part of this same tradition. But they have been induced, by their pretraining, to place this particular task outside it. This task, instead, goes with the shoe-tying or tooth-brushing tradition. If, however, they were required to engage in this task for eight or more hours a day, day after day, week after week, year after year, the effect might be considerably more dramatic. And if everyone around them were engaged in a similarly repetitive activity, the effect might be more dramatic still. For instead of simply failing to locate this task in the problem-solving tradition, that tradition might erode and disappear all together. Adam Smith, of all people, captured this possibility most forcefully. Recall his concern about the side-effects of factory work:

> The man whose life is spent in performing a few simple operations . . . has no occasion to exert his understanding, or to exercise his invention in finding out expedients for difficulties which never occur. He naturally loses, therefore, the habit of such exertion and generally becomes as stupid and ignorant as it is possible for a human creature to become.

Exactly so.

Economic Imperialism: Individual Goods and Social Bads

Market-like arrangements . . . reduce the need for compassion, patriotism, brotherly love, and cultural solidarity as motivating forces behind social improvement. . . . Harnessing the "base" motive of material self-interest to promote the common good is perhaps the most important social invention mankind has achieved.

C. L. SCHULTZE

*W*e have now explored the core principles of the disciplines of economics, sociobiology, and behavior theory and have seen how they share a commitment to developing a scientific conception of self-interested, reinforcement- and fitness-maximizing, rational, economic man. We have also identified some of the limits of that conception. People do not always conform to the standards of economic rationality. Furthermore, how closely people approximate economic rationality may be a matter of historical and cultural circumstance. Under some social conditions, people may look much more like rational, economic men than they do under others.

Thus, while the account of human nature shared by the three disciplines under discussion may provide a largely accurate description of the way things are, the reason for this accuracy is that these disciplines have contributed to, and helped justify, the conditions that foster the pursuit of economic self-interest to the exclusion of almost all else. The experimental demonstration described at the end of the last chapter is an illustration of this process. It shows how one can turn people into pigeons; scientists into factory workers. It shows how people *and* pigeons can be turned into economic men.

The significance of all this is that economics, sociobiology, and behavior theory have not yet delivered the goods as sciences. They have not revealed the eternal laws that govern human nature. Indeed, they may never reveal those laws. For if what is "human nature" depends on the social institutions in which people are immersed, and if those institutions can be altered—even transformed—by human action, then what people are—what human nature is—will depend upon what people make themselves and their society.

Even if all of the arguments that have been presented thus far are convincing, what difference does it make? So what if these disciplines haven't yet earned the right to be called "sciences" in the way that physics has? So what if their principles aren't true of all societies at all times? They certainly appear to be true of *this* society, at *this* time. They certainly help us make sense of the world we inhabit. And even if they never become real science, they have already made an enormous contribution to human welfare. Perhaps people don't have to be rational economic men, but the evidence suggests that if they are, previously unimagined prosperity and social order await them. Perhaps we should regard these disciplines as providing not eternal, scientific truths but a powerful technology—a psychotechnology. People don't have to have airplanes, but if they do, they can travel long distances very fast. People don't have to have computers, but if they do, they can manage enormous quantities of information at high speed. People don't have to be rational economic men, but if they are, they will attain a spectacular level of material well-being. In short, perhaps the psychotechnology of economic rationality has made possible the best of all actual societies, if not the best of all possible ones.

In this chapter and the next, I will try to show how rational economic men, in pursuit of individual "goods," can generate social, or collective, "bads." These "bads" are outcomes that virtually everyone would agree are undesirable; they are outcomes that no one wants. Yet, they seem to be outcomes that we can't help but get by continuing to pursue rational self-interest.

Showing this should not be enough to convince anyone that the pursuit of economic rationality is a mistake. Perhaps these bads are inevitable; perhaps no system of social organization could prevent them.

Or perhaps these bads, while not inevitable, are far outweighed by the goods of economic rationality. To be convincing, I will have to show that these collective bads could be avoided if people lived their lives differently. I will also have to show that they *should* be avoided, that collective bads more than compensate for individual goods. Accomplishing this will require a number of steps. First, we will examine some examples in which the truly rational pursuit of self-interest leads to outcomes that are bad for society as a whole. Second, we will examine how these bad outcomes could be avoided. It will turn out that preventing collective bads requires that one act *non-economically*, that one submerge one's individual interests for the public good. This, in turn, will depend upon people acting as moral rather than as economic men, on their having a strong sense of right and wrong, and on their willingness to act on that basis rather than on the basis of economic self-interest.

But even showing this will not be enough. Remember the economist's claim that "economic man" is not all of man. All that most economists argue is that the economic portion of human nature is important and autonomous of the other portions, whatever they may be. Thus, "economic man" and "moral man" could coexist, each operating in its appropriate domain. Economic man could get people the material prosperity that comes from exchange in a free market, while moral man kept people public-spirited enough to prevent the occurrence of collective bads. What finally needs to be shown is that economic man may be destroying moral man, and that without moral man, economic man may cease to exist. This I will try to show in two different ways. First, the pursuit of economic interests can spill over into domains of life that most people do not regard as economic. A consequence of these spillover effects is to bring more and more aspects of life into the economic sphere. This phenomenon of *economic imperialism* implies that, as people continue to act as rational economic agents, "moral man" diminishes in stature. Society may find that when it needs to call on moral man to prevent some collective bad, he will no longer be there to respond.

Second, rather than being autonomous, the behavior of economic man in the market depends upon the simultaneous presence of moral man, at least if the market is to cater efficiently to people's wants and needs. Thus, as economic imperialism crowds out the moral side of human nature, it undermines the smooth functioning of the market. This is why the goods of economic rationality cannot, in the end, outweigh the bads. Not only are the bads bad, but they also make the market less able to deliver the goods.

In a sense, economic imperialism is an inevitable result of the development of economics, behavior theory, and evolutionary biology as sciences. One of the principal goals of these disciplines is to resolve some of the ambiguity about dividing the worlds of is and ought by locating a wide range of human actions squarely in the domain of facts—of science. If the pursuit of self-interest, be it preference (economics), fitness (biology), or reinforcement (behavior theory), is a law of human nature, then arguing about whether people *ought* to be this way loses much of its force. So every step in the development of these disciplines as sciences is an appropriation of a piece of moral man by scientific, economic man. And with that appropriation comes another step in the direction of economic imperialism.

We may look around us at the world that free-market exchange has made and choose to purchase, in the market of ideas, the picture of human nature that these disciplines are selling. But before making that decision, we should at least have an accurate assessment of the true costs of that picture. So let us examine the price, in collective bads, of buying into the single-minded pursuit of individual goods.

Games People Play

Less than a decade ago, the United States was caught in the grip of massive inflation, a rise in the cost of living of more than 10 percent per year. While citizens of other nations have grown accustomed to rates of inflation that are much higher than this, for Americans, "dougle-digit" inflation was quite a shock. Older people who had retired on fixed incomes they thought more than adequate to see to their needs watched horrified as inflation ate into their cash reserves. Younger people just starting out in careers were stunned to find that each salary increase left them a little father behind than before. And people in mid-career were in despair over where they would find the fifty-thousand dollars needed to pay for their children's college education. The high inflation did not last very long, but it had effects that continue in the present and will persist long into the future. Eroded nest eggs will never be replaced. Furthermore, the fear that if inflation happened once it can happen again leaves everyone worried that no matter how much he has, it may not be enough. It may take an inflation-free period of several generations to convince people that they can plan for their futures with some confidence about what the economic world will be like.

What produces inflation? No doubt, the causes of inflation are many and varied. There is surely no widely agreed upon, simple answer to this question. However, there are some processes that contribute to inflation that illustrate how individual, economic rationality can lead to results that no one wants.

Think about how someone would act as a rational economic agent in a time of modest, but not insignificant inflation. She gets her paycheck and first takes care of the essentials: rent, utilities, basic foodstuffs, and so on. What does she do with the money that remains? She can save it or spend it. If prices are going up at a rate that equals or exceeds the going rate of interest, saving money seems foolish. While it appreciates at an annual rate of, say, 6 percent, prices will be appreciating at an annual rate of, say, 7 or 8 percent. As a result, the things she can just barely afford to buy now will exceed her resources in the future. The sensible thing to do is buy, buy, buy—even things that she doesn't need now but may need next year. And since she is not the only clever person on the block, everyone will decide, for similar reasons, to spend everything they earn. The result is excess demand for goods, what is known as "demand-pull" inflation. Everyone's rush to the marketplace to purchase things and beat inflation serves to make inflation considerably worse. Sellers can raise their prices with impunity, knowing that there will be buyers for their merchandise at almost any price. Furthermore, workers agitate on the job for wage increases that at least match the inflation rate. By doing this, they increase the cost of whatever goods or services they produce, what is known as "cost-push" inflation. They need higher wages to compensate for higher prices. But their wages only push prices still higher, and the inflationary spiral is underway.

It is not hard to prevent these influences on inflation from occurring. All that is required is that individuals exercise restraint, both in wage demands and in purchasing decisions. If everyone accepts smaller than optimal wage increases and defers some consumption by saving, both demand-pull and cost-push inflation will be ameliorated and prices will settle down. So it follows, then, that it is in everyone's interest to be restrained. Right? Wrong. What is really in any individual's interest is for everyone *but* him to exercise restraint. He can then demand and receive a big wage increase and use it to buy goods before his money is ravaged by inflation. He can then capitalize as a free rider on the restraint exercised by others. And this is true for each and every person who is earning and spending money in the society. If everyone but him shows restraint, he gains by not showing it; if no one shows restraint, he at least stays even with everyone else by not showing it. So no matter what everyone else decides to do, his own interests are best served by a distinct lack of restraint. And because this is true for everyone, inflationary spiral inevitably follows. A collective result no one wants follows from the strategy that all rational agents should pursue.

This example has many of the characteristics of the prisoner's dilemma, discussed in Chapter 7. The lesson of the prisoner's dilemma is that the result that is *collectively* optimal is not *individually* accessible. Agents making independent decisions will be led inexorably to an outcome that neither one wants. And so it is with inflation. As long as people act independently, the collective result is suboptimal.

The important thing to realize about the prisoner's dilemma is that one needn't be mean-spirited, greedy, or selfish to be caught up in it. As long as the possibility exists that someone else will be mean-spirited and selfish, one is compelled to act that way in self-defense. If everyone is bidding up wages and prices, inflation will occur no matter what any single individual does, and anyone who exercises restraint, out of some moral principle, will be clobbered. The very structure of the prisoner's dilemma requires that people pay a steep price for high-mindedness.

Just how common is the prisoner's dilemma-type situation? Economist Fred Hirsch argues that prisoner's dilemma-type situations are extremely common. They occur in situations in which people's ability to consume depends upon the magnitude of their resources in relation to the resources of others, in short, on their *relative* economic position. For example, no matter how wealthy society becomes as a whole, not everyone will be able to own a secluded acre of land at the seashore. As more people want one and find one almost within their material grasp, those with still greater resources will simply bid the prices up. Similarly, not everyone can have the most interesting jobs. Not everyone can go to the best college or belong to the best country club. Not everyone can have servants. Goods like these Hirsch calls *positional,* just because not everyone can have them. How likely anyone is to get what he wants depends upon his economic position relative to the position of others. No matter how much money a person has—no matter how much his real (inflation-adjusted) income grows—if everyone else has at least as much, his chances of enjoying these positional goods are slim. Some people, like economist Lester Thurow in his book *The Zero-Sum Society,* have suggested that positional consumption helps turns an economy into a "zero-sum game." As in poker, everything an individual gains he gains at some other individual's expense. Absolute improvements in prosperity are overshadowed by one's relative position on the economic ladder.

We might all agree that everyone would be better off if there were less competition for good grades in school, less competition for good jobs, less striving to improve one's social position, less imposition of absurdly high entry-level job requirements, and so on. We would all be better off if people were oriented to holding their social position and being satisfied with what they have instead of always striving to have more. So much competitive striving is wasteful after all. Students work to get good grades even when they have no interest in their studies. Many go to college, not because they are interested in further education, but to defend their position against everyone else who goes to college. People seek job advancement even when they are happy with the jobs they already have. If only we would all agree to stop. But we can't, because for each individual the best situation is if everyone stops but him. It's like being in a crowded football stadium, watching the crucial play. A spectator several rows in front stands up to get a better view. A chain reaction follows. Soon everyone is standing, just to be able to see as well as before. Everyone is standing rather than sitting, but no one's position has improved. And if someone, unilaterally and resolutely, refuses to stand, he might just as well not be at the game at all. Again, when the goods people pursue are positional, they can't help but be in the race. To choose not to run is to lose. Nonparticipation is not an option.

The positional character of so many economic pursuits points to a problem with the economists' claim that the economic piece of human nature is autonomous of the other pieces. For the moment, divide human nature into two parts: the economic, self-interested part, and the moral, public-spirited part. The economist would have us believe that these parts can coexist within people peacefully and independently, each guided by its own rules of conduct—preference maximization in the economic domain, and adherence to some ethical code in the moral domain. But the fact that pursuing individual goods can yield collective bads makes any decision to do the right, public-spirited thing an *economic* decision. If we decide to defer consumption and wage demands because inflation is a collective bad, we are making an economic sacrifice. We are following an irrational economic strategy. The economist might say that moral action is a good—a luxury good perhaps—and if people decide to purchase it, they do so at the expense of other goods they might have if they decided not to worry about acting morally. At least in modern society, if not in general, the economic and moral spheres of action cannot be walled off from each other.

This point can be illustrated concretely with an example discussed by Thomas Schelling, in his book *Micromotives and Macrobehavior*. Consider one of the most significant economic acts that most people perform—the purchase of a house. Economically, this is no small matter. In buying a house, people incur a huge debt. Often, they buy a house that they can't quite afford, counting on tax advantages and appreciation to keep them above water. So considerations like mortgage terms, potential resale value, likely maintenance expenses, and the like will have a lot to do with which house they choose. However, this major economic act has many noneconomic characteristics that ride along with it. For in purchasing a house, people also purchase neighbors, and schools for their children. And they can't help but do this. The economic aspects of house buying can't be segregated from the noneconomic ones. These two aspects of housing decisions impinge on each other.

Suppose a family has a moral commitment to racial integration, believing it is important that neighborhoods be integrated in general and important for their own children that they grow up in a multiracial environment. Unfortunately, the resale value of houses in multiracial neighborhoods is much less certain than it is in segregated neighborhoods. So the purchase of multiracial neighbors entails considerable economic risk. Or suppose that a family purchases a house in an all-white neighborhood. After they have settled in, black people buy a house down the street. As committed supporters of integration, they are pleased.

Then, one of their white neighbors puts his house up for sale. He has nothing against blacks, but he knows that other people do, and he is worried that the value of houses in the neighborhood is about to diminish. His house is purchased by another black family. Now others of their white neighbors start to worry about the value of their houses. Other houses are put up for sale. Because there are so many of them, they fetch lower prices than one might have expected. Some of these are purchased by blacks. On the one hand, the family is happy that their neighborhood is becoming truly integrated. On the other hand, each step in the direction of integration is a threat to the value of their most substantial investment. Can they *afford* not to sell their house as more and more of their white neighbors decide to sell their houses? Eventually, they can no longer pay the price that morality costs, and they too sell their house, for less than they could have two years ago, and buy a new one in another white neighborhood. The neighborhood has tipped, from all-white to all-black, as a result of the rational, economic decisions of a set of unprejudiced individuals. It seems that economic man and moral man just can't help but affect each other.

The Problem of the Commons

Imagine a small village of craftspeople in the eighteenth century. Each family has a house with a small plot of land for raising vegetables. In addition, there is a large, common area used by all the villagers to graze their livestock. Each villager has a few cows that provide the family with dairy products. The common is large enough to support the entire village. Then the village begins to grow. Families get larger and procure an extra cow. New families move in. Suddenly, the common is threatened; it is being overgrazed. Grass is being trampled. It is being consumed so fast that there is not enough time for it to replenish itself before rains erode the topsoil. Each cow no longer has quite enough to eat and thus yields less milk than it did before. If the overuse of the common continues, there will be a slow but sure decrease in the number of animals it can support, until finally it becomes useless for grazing.

How can the overuse be stopped? Consider the issue from the point of view of an individual villager. He needs the milk from his two or three cows, and even if they give him less than they did before, less is better than nothing. Besides, how much difference will it make if he alone shows some restraint in his use of the common? Indeed, his temptation might be to add another cow, since each of the ones he already has is producing less than before. The slow decrease in overall production of dairy products in the village has little impact on him, especially in comparison to how he would be affected if he stopped using the common altogether. But of course, this villager's interest, that he continue to use the common as before, is shared by all the villagers. The result is that none of them modifies their behavior, and the common is destroyed.

This is the "commons problem," a problem made prominent recently by Garrett Hardin, but known to economists in one form or another for many years. The problem is one in which the interests of the collective, and even the long-term interests of each individual in the collective, are best served by the exercise of restraint, but only if everyone cooperates in showing that restraint. In the absence of such cooperation—if some individuals are likely to take advantage of the restraint shown by others—then it is in no one's interest to show restraint.

Technically, the commons problem exists when only those who use the common are affected by its overuse and the cost of overusing it is in the same currency as the benefits, in this example, milk yield from cows. Examples that meet these technical requirements include the overharvesting of trees by lumber companies, the overplanting of land by farmers, the overdevelopment of suburban communities, the extraction of petroleum from a common pool by oil companies, and the overcrowding of highways, beaches, libraries, and other public facilities so as to make whatever benefits its users derive from those facilities vanishingly small.

However, if we relax these requirements a little, the examples proliferate endlessly. Pollution of water by toxic wastes, pollution of the environment by noise and litter, overpopulation, hoarding of goods during shortages, driving recklessly, using water to keep a green lawn during droughts, keeping the air conditioner on during hot summer days when the community is threatened with a power shortage are all versions of the commons problem. The costs and benefits may not be in the same currency, and some of the costs may spill over to people who are not deriving the benefits. But in all of these examples, what is rational for the individual depends on what other individuals do. In the absence of any assurance that all individuals will exercise restraint for the common good, the rational individual will eschew restraint rather than be exploited by free-riding others.

Consider the suburbanization of large cities. The impetus for moving to the suburbs is presumably to get access to a little clean air, grass, and open space, without being so far away from work that commuting is intolerable. So one moves to a suburb—and so does everyone else. The result is dirty air, crowds, noise, and a commute. So one moves a little further out—and so does everyone else. The crowds are as bad as before, but the commute is worse. One can continue moving further out, until one is no longer in a suburb but in the country. The trouble is that the country isn't the country anymore. At best, it's a moderately crowded suburb. And now, the commute from crowded suburb to crowded city is ninety minutes each way.

This happens, in part, because what *anyone* can have, not *everyone* can have. It is not just one person who gets the bright idea of combining country charm with proximity to the city. Because so many people get the same idea, country charm erodes. There are ways, of course, of preventing suburbs from becoming as congested as the cities to which they were meant to be the antidote. Zoning regulations are most common. If only single-family dwellings are allowed and lots must be sold that are at least a half acre in size, overcrowding can be prevented. What will occur instead is a bidding war, in which only the very rich will be able to have their desires satisfied. So for most people, the end result of the pursuit of suburban tranquility will be frustration. Either they won't be able to afford it, or it won't be tranquil. Like the mass of people all of whom are standing on tiptoe, at best people will end up expending extra effort (commuting time) just to hold their position.

There is an important and quite general feature to the suburbanization process that should be pointed out—what Schelling has called "the tyranny of small decisions." When contemplating moving to the suburbs, the choice one is faced with is this: crowds, noise, and no commute vs. quiet, space, and commute. One chooses the latter. But because so many others do the same, the result is that what turns out ultimately to have been chosen is crowds, noise, and commute over crowds, noise, and no commute. Anyone who had been faced with *that* choice at the beginning would have stayed where he was.

This happens all the time. The choices people are confronted with at the moment are very different from what they would be if people knew that everyone else was confronted with the same choices and was likely to make the same decisions. For example, imagine two bookstores, side by side. One is a discount chain with little selection but good prices. The other is an old-fashioned bookstore with seemingly everything in stock and a knowledgeable proprietor but list prices. A current best seller is available at both stores, but it costs fifteen dollars at the old-fashioned bookstore and ten dollars at the discount chain. A rational purchaser will, of course, go to the chain. So will everyone else. The old-fashioned bookstore will go out of business. Faced with the choice between spending a few extra dollars and keeping the old-fashioned store alive, or saving money and putting it out of business, would everyone still have chosen to save the money? The logic of free-market individualism dictates that people will often get long-term, collective outcomes they don't want if each of them, as an individual, pursues the economic ends he does want. And so it is with the transformation of suburbs into inconvenient cities.

It seems that part of what people purchase when they buy into the

free-market, individualistic organization of society is a host of collective prisoner's dilemmas and commons problems. Is the free market
worth this price? Could people avoid these problems if they organized
society in some other way? Could people avoid these problems if they
just patched some restraints onto the free-market system? Or are these
problems that people simply have to live with no matter how society
is organized?

In recent years, people have suggested three different types of
answers to these questions. One of them is voluntarist. Educate the
populace about the dangers and social costs of pollution, reckless
driving, wanton use of energy and public lands, overpopulation,
smoking, drinking, and the like and exhort them to do their duty as
citizens and exercise restraint. This strategy appeals not to economic
man but to moral man. It tells people what the right thing to do is
and shows that if they all do the right thing society will be better off.

The other two strategies appeal to economic man directly. One of
them offers incentives for "good behavior." Nonpolluters and energy
conservers get tax breaks. Nonsmokers pay less for life insurance,
and safe drivers pay less for car insurance. The other provides charges,
or penalties for "bad behavior." Polluters pay steep fines. Purchasers
of inefficient automobiles pay a luxury tax. Tobacco and alcohol are
taxed, to help defray the social (medical) costs of tobacco and alcohol
use. Permits are sold to regulate the use of parks and beaches. Access
to highways during rush hours is restricted to force people to car pool.
Restraints like these are designed to change the economics of each of
the relevant situations, to make individual, economic self-interest line
up with collective interest. One can choose not to exercise restraint,
but only at a price. And the price will be high enough either to induce
restraint or to compensate society for profligacy.

Until recently, these restraints proliferated; every month, it seemed,
another social cost was identified and another set of regulations designed
to mitigate that cost was created. Free-market ideologues railed against
these restraints on grounds of principle, and of late they have been
in a position to undo them. Business people complained about them
because they made it difficult to operate profitably or to react quickly
to ever-changing market opportunities. And even many public-spirited,
non-market-oriented citizens were unhappy with them—because they
didn't work.

Why didn't they work? Why can't a set of laws that provide rewards
for public-spirited actions and punishments for self-interested, harmful ones succeed in achieving restraint? After all, we live in a society

of laws, and our ability to enforce those laws is good enough so that most people obey most laws most of the time. We live in a society in which people make legal or contractual commitments to one another, and our ability to enforce those commitments is good enough so that most people honor most contracts most of the time. Why can't we make people toe the line and act for the common good? The reason why is that our ability to enforce laws and contractual commitments depends critically on the fact that most people honor them, not out of fear of being caught in transgression, but because it's the right thing to do. Suppose that everyone decided to do whatever he thought he could get away with, no matter what the law said. People would cheat on taxes, go through red lights, sign contracts and then ignore them, refuse to pay bills, and the like. How many people would get caught? And if they were caught, how long would it take for their transgressions to be adjudicated in the courts? Already our legal system is so jammed up that civil suits take years and years to be decided. Already our jails are so crowded that only people accused of the most serious crimes are incarcerated.

And this is in a society of predominantly law-abiding citizens. As more people decided to ignore the law, the chances that any of them would get caught and punished would get lower and lower. If everyone decided to ignore the law, our system of taxes, fines, and other punishments would quickly become paralyzed and useless. It's a kind of "critical mass" phenomenon. If one person attempts to cross a busy intersection against the light, there is no chance that cars will stop to let him pass. But if dozens of people do it together, the light becomes irrelevant; traffic stops. If our system of legal constraints works, it is only because a critical mass of people has not yet decided to disregard laws and contracts.

Furthermore, even if we could enforce all laws and contracts, there would be big trouble if we had to depend exclusively on them to get people to act appropriately. One simply can't put everything in a contract. To a large extent, the smooth functioning of the economic and social system depends not on contracts but on a measure of good will— on adhering to the spirit rather than the letter of the law. If evidence of the limited value of contracts is needed, consider how some of the most effective devices workers have for exerting control over management involve not going on strike but honoring their contracts—to the letter. Such job actions—known as "working to rule"—paralyze productivity. For as soon as anything unanticipated in the contract comes up (as it inevitably will), workers substitute contractual obligation for judgment and do nothing. We count on workers to enter into agreements with their employers in good faith, to understand the point of their work activity, and to use their discretion in pursuing that point when unforeseen difficulties arise. Our system of legal and contractual constraint works precisely because for most people it does not have to be applied.

So economic incentives and disincentives will not do the job, at least as long as people depend on them exclusively. Happily, such economic devices are not the only ones at our disposal. Recall the first strategy mentioned above: appeal to voluntary restraint as an act of public virtue. This strategy would be relatively cost-free. Enforcement would not be a problem, since compliance would be voluntary. But could it work? Clearly, it could not work as long as people were exclusively economic men. Public virtue is for suckers. Economic rationality absolutely demands that one free-ride if he can get away with it. But people are not exclusively economic men. The trick, then, seems to be to appeal to moral man to solve a social problem created by economic man.

Moral Restraints on Economic Activity

That the effectiveness of economic man depends upon the simultaneous existence of moral man is not news. Adam Smith said it two hundred years ago:

> All members of human society stand in need of each others' assistance . . . where the necessary assistance is reciprocally afforded from love, from gratitude, from friendship and esteem, the society flourishes and is happy. All the different members of it are bound together by the agreeable bonds of love and affection. . . . Society, however, can not subsist among those who are at all times ready to hurt and injure one another.

Smith's views were echoed by philosopher John Stuart Mill. In his book, *On Utilitarianism,* Mill argued the need for sympathy explicitly on the grounds that externally imposed restraints were not sufficient to secure social order. For Smith, this need for love, friendship, and esteem was not a problem for the market system, because he thought that "natural sympathy" was as much a part of human nature as was the pursuit of self-interest. People wouldn't do absolutely anything to secure their interests. What they would do was constrained by their concerns for others.

The modern economist might acknowledge both the need for, and the presence of, sympathy, but suggest that it can be viewed as an economic good. If doing harm to someone else makes a person feel bad, then it is in that person's interest not to do harm. Behaving "morally" is simply the way to maximize his own, personal welfare. If the satisfaction a person gains by using less of the common than he might, and making his neighbor happy as a result, is greater than the satisfaction he loses in lowered milk yield, then restraint is the rational, *economic* decision. No appeal to morality—no concern for the public good—is required.

Thanks to the development of both sociobiology and behavior theory, it has become easier to regard acts of "love" or "friendship" as no less economic than any other activity; the goods involved are just social and psychological, rather than material. So in deciding whether to exercise restraint in the interests of the public good, people engage in *economic* calculations: how much material welfare will be gained by refusing to exercise restraint; what will it cost to be caught; how good will showing restraint make other people feel and how good will that make me feel; how many units of sympathy is that extra quart of milk worth? If, after engaging in these calculations, the bottom line says be restrained, people are restrained; if not, they are not.

Just as a system of incentives and penalties can't ensure the public good, neither can acting out of sympathy, that is, acting because making others feel good makes oneself feel good. Sympathy might work in a small community, in which everyone knows everyone and the effects of one's actions on others are discernible. But it quickly breaks down as small communities become large cities, as buyers and sellers in the market become virtually anonymous, indeed, as many transactions occur by long distance.

Under these conditions, in which increasing numbers of people live, the pull of sympathy fades. Who is it that feels bad if someone neglects to pay a credit card bill? Who is the citrus farmer hurting if he charges as much as he can for the oranges that are about to be shipped halfway across the country? It is the very anonymity of the market that ensures its efficiency, that ensures that price will be governed by supply and demand, and *only* by supply and demand. But a price of anonymity is reduced sympathy. If we are to appeal to anything to induce people to act for the public good, it had better be to something other than sympathy.

What could that other "something" be? Economist Amartya Sen has written about the economic, self-interested notion of sympathy.

He acknowledges that "sympathy as self-interest" surely exists. But he further argues that there is another source of action for the common good that the notion of sympathy as self-interest can't encompass, indeed, that sometimes leads to actions in direct violation of self-interest. He calls this other source of public-spirited action "commitment" and suggests that it cannot be incorporated within the economic framework.

To act out of commitment is to do what one thinks is right, what will promote the public welfare, quite apart from whether it promotes one's own. It is to act out of a sense of responsibility as a citizen. Acts of commitment include voting in large general elections. No matter how small the cost of voting to a single individual, in time or inconvenience, it will be larger than the benefit. One vote simply won't make a difference. Yet people vote, or at least many people do. Acts of commitment include coming to the aid of a stranger in distress, always at a cost of time, and sometimes at considerable personal risk. Sociobiologists explain such altruistic acts toward strangers, that is, nonrelatives, by appeal to the concept of "reciprocal altruism," a kind of "you scratch my back and I'll scratch yours." But in the large, anonymous cities in which many people live, the likelihood that acts of altruism will ever actually be reciprocated by their recipients is miniscule indeed. Acts of commitment include doing one's job to the best of one's ability—going beyond the terms of the contract, even if no one is watching and there is nothing to be gained from it. They include refusing to charge what the traffic will bear for necessities during times of shortage, refusing to capitalize on fortuitous circumstances at the expense of others.

Acts of commitment like this occur routinely. They are what holds society together. But they are a problem for economics. As Sen says: "Commitment . . . drives a wedge between personal choice and personal welfare, and much of traditional economic theory relies on the identity of the two." And he continues:

> The economic theory of utility . . . is sometimes criticized for having too much structure; human beings are alleged to be "simpler" in reality . . . precisely the opposite seems to be the case: traditional theory has *too little* structure. A person is given *one* preference ordering, and as and when the need arises, this is supposed to reflect his interests, represent his welfare, summarize his idea of what should be done, and describe his actual choices and behavior. Can one preference ordering do all these things? A person thus described may be "rational" in the limited sense of revealing no inconsistencies in his choice behavior, but

if he has no use for this distinction between quite different concepts, he must be a bit of a fool. The *purely* economic man is indeed close to being a social moron.

It must be noticed that commitment cannot be viewed as a piece of moral man, independent of economic man. For acts of commitment will have an impact on economic activities. Indeed, to act out of commitment is, almost by definition, to forgo economic considerations and choose on some other basis. The existence of commitment puts economic decision making in a whole new light. True, when making economic decisions, people will presumably choose that alternative that maximizes preferences. But before they can do this, they have to make another choice. They have to choose to make an economic decision, based on self-interest, as against, say, a moral one that is based on commitment. And we can't say whether choosing to make an economic decision or not in any given situation is "rational" (that is, preference or utility maximizing) since there is no common scale on which moral and economic decisions can be compared. Is it rational or not to refuse to profiteer in times of shortage? Is it rational or not to help out a stranger in distress? Economics does not have the answer.

Economic Imperialism and the Undermining of Commitment

The importance of commitment—of moral man—to the securing of the common good raises what is perhaps the most serious problem with the pursuit of self-interest, as envisioned by the economist. Despite Adam Smith's recognition that economic man was just a part of man, and despite his view that economic man and moral man could—indeed had to—coexist, there is ample evidence from modern social and economic life that economic man just won't stay in his place. He seems bent on encroaching on moral man, determined to appropriate more and more aspects of daily life to his own domain. The spread of economic considerations to previously noneconomic aspects of life has been called *economic imperialism.* It is time now to examine economic imperialism.

Economist Fred Hirsch identifies many aspects of modern life that once were largely independent of economic considerations but are now becoming increasingly pervaded by them. One example he discusses is education. With increasing competition among members of society for good jobs, employers keep erecting new hurdles that must be jumped before job entry is possible. First high school degrees were required, then college degrees, then special training programs, then master's degrees, then doctoral degrees, then doctoral degrees at only a handful of select, certified institutions. The training one receives in these various programs may or may not be relevant to the requirements of the job; relevant training is really not the essential point. What is critical is to make the path to the job arduous enough so that only a few dedicated souls will embark on it.

These hurdles have a profound effect on the way people view education. With education so closely tied to job entry, and job training, it becomes an "investment" in one's future. The money spent on school is expected to be returned, *in kind,* and with interest, later on. One can put a dollar value on a college degree by surveying the wages paid on the jobs to which it gives access. It is easy to imagine people engaged in the following kind of calculations: A degree from Harvard will cost sixty thousand dollars. If we took that money and invested it and entered the job market four years earlier than we otherwise would, would the interest on investment coupled with the extra years of earning power compensate for the high-paying jobs forgone? Or perhaps the calculations might go like this: Harvard will cost sixty thousand dollars, while the state university will cost twenty thousand dollars. Will the job opportunities provided by the Harvard degree pay back the extra forty thousand dollars invested?

It is easy to see how thinking about education in these terms—as an economic investment—can affect what people want out of education and thus how they evaluate what they get. If enough people assessed their educations in these terms, what actually went on in the college classroom would surely change. Colleges and universities would have to be sensitive to market demand; they would have to provide what students wanted, or the students would go elsewhere. The goal of education would shift from creating well-informed, sensitive, and enlightened citizens to creating skilled workers. This is an example of economic imperialism. To the claim that one can't put a dollar value on having an educated citizenry comes the reply, of course one can. One simply looks at how much extra salary the education makes possible. Extra salary becomes the yardstick for evaluating the effectiveness of an educational institution. Before long, the institution changes what it does, so that the creation of extra salary potential becomes the goal itself, instead of just a measuring stick.

Another example of economic imperialism is that everyday social relations—as friends, neighbors, spouses, and parents—are taking on an economic component. In part, this comes from a little-noticed aspect of consumption: it takes time. If one only has money for the essentials of life, finding the time in which to consume them is not an issue. But if one has money for stereos, video recorders, dinners in nice restaurants, the theater, and vacations, one must find the time to decide which stereo, restaurant, play, or resort to partake of. In addition, one must find the time actually to partake of it. Dinner at home is an hour; dinner out is an evening. Listening to records takes time. Recording a favorite television show doesn't, but actually watching the recording does. No matter how rich a person is, time is a resource that cannot be increased; there are only twenty-four hours in a day.

The pressure for time to consume has real costs. It produces what Hirsch calls "the economics of bad neighbors." Time spent being sociable is time taken away from consumption. Chatting over the backyard fence or helping a neighbor cut down a tree are actions taken at a cost of using the video recorder or going into town for a nice dinner. Whether we like it or not, the decision to be sociable becomes an economic decision, another example of the spread of economic considerations to traditionally noneconomic domains. Many people have experienced how much harder it has become to find time to spend a quiet evening sipping beer and chatting with a few friends. It is becoming increasingly rare for such occasions to develop spontaneously; they must be planned days, or even weeks, in advance. And of course it seems ludicrous to "plan" an evening of casual conversation. So instead it becomes a dinner party. This in turn only adds to the time pressure, since now food must be purchased and an impressive meal must be prepared. One of the virtues of having limited resources is that one often can't afford to do much more than have an evening of casual conversation over beer.

This "economizing" of social relations can, in the long run, have profound effects. The strength of people's commitment to the public good may derive in part from the range and depth of their personal, social relations. As these diminish, for lack of time, commitment may diminish with them. Decrease in commitment, then, is a cost of pressure to consume that is not reflected in standard economic calculations. Indeed, even if one's "public-spirited" actions have always been economically motivated, a general decrease in sociability is likely to change the economic incentives to be a good citizen. Suppose that the good turns one does have been motivated only by the expectation that they will one day be reciprocated, and nothing more. As one's network of social relations shrinks, the likelihood of reciprocation shrinks with it. With each decrease in social connectedness, there comes a corresponding decrease in the possibility of concerted action to solve the various commons problems all people face.

If many of our social activities are increasingly taking on an economic aspect, accurate assessment of collective productivity will require some way of measuring the economics of social goods. And some economists have suggested that the best way to do this is to take these domains out of the economic closet and treat them explicitly as part of economic life. We should enlarge the consumer's potential market basket to include not just stereos, vacations, and dinners out, but time spent chatting with friends as well. Even if we don't literally put dollar values on the time people spend with friends, there ought to be some way to estimate the dollar value equivalents of social activities, perhaps by examining how many things that do have dollar value people are willing to trade for time spent with friends. Though this move may seem a bit contrived, it will at least have the virtue of revealing costs and benefits that were previously hidden from economic analysis. Making this move is a major step in the direction of economic imperialism; it represents the commercialization of social relations.

The Commercialization of Social Relations

In the economic world, people get what they pay for. Certainly, they get nothing more, and vigilance is required to see that they don't get less. People are not in business for their health, after all. So what happens when the social world gets commercialized? Presumably, people start getting only what they pay for in social relations as well as economic ones. Now in the economic world, people are prepared to operate on this assumption. Products come with explicit guarantees, services are provided in accordance with detailed and specific contracts. People enter into exchanges with their eyes open, expecting, and guarding against, the worst. They are not so prepared in the social world, or at least have not been until recently. People assume that friends, lovers, families, doctors, and teachers will act with good will, doing, insofar as is possible, what is best for them. As a result, they ask no guarantees and write no contracts. People trust that part of what it means to be a spouse, lover, parent, doctor, or teacher ensures that people close to them will behave honorably, truthfully, courageously, and dutifully in social interactions.

As social relations become commercialized, however, this assumption grows more and more suspect. Increasingly, people feel the need to have things written down in contracts. Increasingly, they feel the need to be able to hold others legally accountable—whether doctors, lawyers, teachers, or even friends or lovers—to have a club to wield to ensure that they are getting what they pay for out of their social relations. Many people have come to expect the worst from marriage. Some radical feminists argue that marriage is nothing but a combination of legalized prostitution and indentured servitude, in which men take helpless women for all they are worth. To protect themselves, some men and women insist on marriage contracts that are not just metaphors. "For better or for worse, in sickness and in health, till death do us part" is replaced by a detailed specification of household and sexual obligation and entitlement and asset ownership and control, to be reevaluated by both parties after a fixed term.

The economist might argue that this shift from a dependency on what is implicit in various social relations to what is explicit and contractual is merely a recognition of cold, hard reality. And these tendencies to put things down on paper are a legitimate attempt to redress real grievances. It is an extension of the consumer protection movement to social domains, to protect people from doctors who are callous or paternalistic, teachers who are indifferent, and spouses who are out to dominate and exploit.

But what the economist who makes this argument fails to realize is that the process of commercialization of social relations affects the product. By treating the services of doctors and teachers as commodities being offered to the wary consumer, we change the way doctors doctor and teachers teach. Doctors practice defensively, doing not what they regard as the best medicine but what they regard as the best hedge against malpractice suits. Medical costs soar, but medical care does not improve. Teachers teach defensively, making sure their students will perform well on whatever tests will be used to evaluate their progress, at the expense of genuine education. Test scores go up, but students are no wiser than before.

There is, in short, a self-fulfilling character to the commercialization of social relations. The more that we treat such relations as economic goods, to be purchased with care, the more they become economic goods about which we must be careful. The more that an assumption of self-interest, rather than commitment, on the part of others governs social relations, the truer that assumption becomes. As Hirsch has said, "The more that is in the contracts the less can be expected without them; the more you write it down, the less is taken—or expected—on trust." And further, "Orgasm as a consumer's right rather rules it out as an ethereal experience." It is the fulfillment of the sociobiologist's vision. We replace the view that the people close to us love us and are deeply concerned for our welfare with the view that they are out only for themselves.

Practices and Their Contamination

To appreciate better the character and significance of economic imperialism requires a clearer idea of what it means for a domain of activity to be noneconomic. If the pursuit of self-interest or utility does not set the goals of a domain, then what does? What is it that participants in noneconomic activities strive for?

This question has been illuminated by Alaisdair MacIntyre, in his book *After Virtue*, in which MacIntyre attempts to reconstruct a moral philosophy. Central to that attempt is the concept of a *practice*. Practices are certain forms of complex and coherent, socially based, cooperative human activities. Among their characteristics are these:

1. *They establish their own standards of excellence and, indeed, are partly defined by those standards.*
2. *They are teleological, that is, goal-directed.* Each practice establishes a set of "goods" or ends that are internal or specific to it and inextricably connected to engaging in the practice itself. In other words, to be engaging in the practice is to be pursuing these internal goods.
3. *They are organic.* In the course of engaging in a practice, people change it, systematically extending both their own powers to achieve its goods and their conception of what its goods are.

Thus practices are established and developing social traditions, traditions that are kept on course by a conception of their purpose that is shared by the practitioners. And most importantly, the goals or purposes of practices are specific or peculiar to them. There is no common denominator of what is good, like utility maximization, by which all practices can be assessed.

Various everyday social relations, like friendship, marriage, and parenting, can all be viewed as practices. They are each complex social activities with long histories. They are each teleological, with ends that continue to evolve. There are people who are good and people who are bad at these practices, although what it means to be good and bad changes as the practices change. Most of us expect good examples of these practices to be characterized by intimacy, trust, care, and concern—attributes that might be regarded as contributing to what Sen called "commitment," and certainly, attributes that are very different from the pursuit of economic self-interest. But since practices evolve, the goods that characterize friendship, marriage, and parenting may be expected to change. Indeed, the commercialization of these relations has precisely this effect. Rather than intimacy, trust, and commitment, the goods of commercialized social relations may be the mutual satisfaction of individual interests.

But if the possibility of change is built into the very notion of a practice, why make such a fuss about the commercialization of social relations? The reason is that this is a very special kind of change. The satisfaction of individual interests is not a good that is specific to the practices of friendship and marriage. It is a good that characterizes all economic activity. Once it also becomes the goal of friendship and marriage, it destroys these activities as distinct, organized, and coherent.

We can see this more clearly by considering another kind of practice in some detail. The collection of activities referred to as "science" is a practice. Sciences are certainly complex, social activities. They establish their own standards of excellence. They have a set of "goods," the pursuit of which partly defines them. And they develop. The goal of science is to discover generalizations that describe and explain the phenomena of nature.

Different scientific disciplines develop traditions that provide guidance as to which generalizations are worth going after, which methods are best suited for going after them, and which standards should be used for determining whether one has succeeded. Now not all people who do what looks like scientific work are engaged in the practice of science. People who do experiments to achieve impressive publication records are not engaged in the practice. The goods they seek—fame, wealth, status, promotion—are not internal to science. Science is just one means to those goods among many. It is certainly true that people who are pursuing such external goods may well do good science; that is, they may well contribute to the development of the practice. But they are not themselves practitioners. And if everyone engaged in science were to start pursuing these external goals, the practice of science would cease to exist. The core of the practice of science—the thread that keeps it going as a coherent and developing activity—lies in the actions of those whose goals are internal to the practice.

And these internal goals are all noneconomic. The scientist does not choose from among a variety of market baskets, each containing some amount of truth and some amount of status and money, in different proportions, the one market basket that maximizes his preferences. One does not bargain away portions of truth for portions of something else, at least not if one is working within the traditions of science. But in the experiment described at the end of the last chapter, some of the experimental subjects did precisely this. They bargained away truth, or more accurately, the best techniques for discovering truth, in return for money. For subjects who were pretrained to perform a particular task for monetary reward, the problem-solving task was not "pure" science. It was an amalgam of truth seeking and money seeking, of doing what will yield a general principle and doing what works. These students struck a compromise between two competing masters when they faced the problem-solving task. Their compromise was not necessarily conscious or deliberate, but it was there nevertheless. They forsook traditional methods of doing good science that their untrained colleagues followed so that they could earn more money.

One doesn't need our laboratory demonstrations to see this compromise of scientific and economic objectives. Economic considerations have been affecting the behavior of real scientists, doing real science, for years, and continue increasingly to do so. The evidence is everywhere. It is not good science to do the same experiment again and again—to repeat what works. Yet with research success, promotion, and the granting of tenure largely determined by rate of publication, many scientists do effectively that. Each experiment is a minor variant on the preceding one, because such mechanical and unimaginative variation is the quickest road to print.

It is not good science to decide what to study on the basis of what people are willing to pay for. Yet government agencies are able to manipulate fields of inquiry by shifting funding from one domain to another. It is not good science to keep one's results a secret, keeping others in the dark, or even intentionally misleading them. Yet, in areas that are "hot," scientists do this routinely, as a way of protecting claims to priority, even at the cost of scientific progress. A dramatic example of this sort of behavior is provided by James Watson in *The Double Helix,* his autobiographical account of the unraveling of the mysteries of DNA.

Above all, it is not good science to lie—to misrepresent results willfully, or to invent results of experiments that were never conducted. Yet, in the last few years, several examples of blatant falsification have been uncovered at major research institutions. This last perversion of science, presumably in the interest of self-aggrandizement, is especially crippling. Science must proceed on the presumption that its practitioners always tell the truth, even if they aren't always successful at finding it. Were this presumption seriously undermined, science would grind to a halt. Either all experiments would have to be repeated by all interested parties, to make sure of the veracity of published reports, or monitors would have to be stationed in all laboratories. In the first case, science would stop being a collective, cooperative, and cumulative enterprise. In the second, we would have to worry about how to make sure that the monitors were telling the truth.

A particularly vivid example of the difficulties for scientific practice that arise when science is penetrated by economics has surfaced in the last few years with the commercialization of microbiology in general and genetic engineering in particular. Recent advances in molecular genetics have opened the way to a mass of potential practical applications. Although none of these applications has as yet been fully developed, and only a small handful are being used commercially at all, the potential for enormous profit is so great that numerous companies have been formed, each in hot pursuit of the brass genetic rings. A number of these companies have gone public, issuing stock, and they have as major stockholders internationally prominent scientists. For the potential investor, the trick is to buy stock in one of these companies *before* they make a major breakthrough—to get in on the ground floor. Once the breakthrough occurs, the stock will skyrocket in price. Thus investors hang on every word about progress toward successful application.

So imagine a scientist of world reknown who is a director of one of these companies, holding, say, one hundred thousand shares of stock. He calls a press conference and announces that his company has just made a significant scientific discovery, bringing application of genetic engineering a major step closer. He can't, of course, disclose the discovery in detail because of the intense competition. Within a week of the press conference, the price of company stock rises thirty dollars a share. The scientist is now three million dollars richer. There is nothing necessarily wrong with reaping a personal profit from a discovery. But notice that in this illustration, the profit comes not from any actual discovery, but from the scientist's *claim* that a discovery has been made. He presented no evidence, but people believed him. Why?

To see why, imagine the president of General Motors calling a press conference to claim that his company is on the verge of producing a car that gets one hundred miles to a gallon of gas. All very hush hush; no evidence. "We've got to worry about the competition. Just trust me." What would happen to the price of GM stock? We can't say for sure, but it is reasonable to suppose that such an announcement would be greeted with substantial skepticism. People know what the interests of the GM president are, and it is perfectly possible that he would make such a grandiose claim to improve the position of his company, whether or not it represented a substantial exaggeration. Such exaggerations, if not out and out misrepresentations, are commonplace. *Caveat emptor.*

What about the scientist? When he makes his announcement, is he wearing his scientist's hat (always tell the truth) or his board of directors hat (always maximize profit)? It is because we assume he is wearing his scientist's hat that we take his announcement so seriously. But he can switch hats—and get very, very rich, very, very fast—long before anyone catches on. What is the poor scientist to do? Just a little exaggeration will make him and his company so much better off—give them capital with which to expand their research efforts. And who knows? The extra research effort may actually produce the breakthrough that the scientist already announced.

Certainly, if scientists start wearing both these hats in increasing numbers and start making exaggerated or distorted claims about discoveries, they will eventually have no more credibility in the public eye than any other business person. But as their credibility is undermined, so will the credibility of science in general. Science will become not a set of practices distinct from the economy and governed by noneconomic considerations but a part of it. The goal of science will be profit maximization, just like any other industry.

Concerned scientists worry that it is just a matter of time before corporate concerns start taking control of the university laboratory, dictating which problems are to be studied, who is to be hired to study them, what information is allowed to be publicly disseminated when, and the like. At the very least, even if they don't dictate the direction of research or encourage distortion, corporate sponsors can be expected to exert substantial control on communication. It does them little good to foot the bill for research if everyone can gain access to its products, through the scientific journals, at the same time they do. Such control already exists in the basic research arms of many commercial ventures like drug and pharmaceutical companies. Scientific manuscripts are carefully screened, to be sure that no important "trade secrets" are being revealed, before they are permitted to be submitted for publication. If practices like this become widespread enough, they will create substantial doubt among scientists about just how accurate and up to date the reports they see in their journals really are.

The economist is a scientist, engaged in the traditional, scientific pursuit of the truth. He wants no part of the undermining of this tradition. He wants his own credibility to be preserved, and he wants the credibility of his discipline to remain intact. But he is in no position to condemn the pursuit of rational self-interest in others. Nor can he count on market competition to keep science pristine. For what makes it pristine, if anything does, is its nonparticipation in the market. So the economist finds himself defending a view of human nature that will systematically and surely erode the very base from which he speaks.

The penetration of economic self-interest into the practice of science is a very significant modern development because of the central place of science in the modern world. But it is not an isolated development. In the world of sports, for example, a similar economic encroachment has been going on for years. At the professional level, it has become a commonplace that sports are just businesses. Organizations buy teams to make money, or even to lose money as long as the losses bring tax advantages. They don't work to promote the pursuit of excellence in the sport; they work to promote the pursuit of profit. Schedules are determined, game sites and times established, and playing conditions arranged not to ensure that athletes will always be able to perform at the limits of their capacity, but to ensure maximal television audiences and the lucrative television contracts that they bring with them. Professional teams relocate at the first sign that there is additional profit to be made elsewhere, or threaten to do so as a way to blackmail their home cities into making concessions on stadium rental and the like.

The athletes themselves, following the example set by their employers, sell themselves to the highest bidder, without concern for team loyalty, continuity, or excellence. At the amateur level, football coaches outearn college presidents, and curricula are established that allow athletes to stay eligible for competition without being constrained to study, so that they can devote all of their time and energy to the sport they have been recruited to compete in. Organized athletics have frequently been defended as builders of the strong character people will need to be upstanding and outstanding citizens of the larger society. The lessons learned on the playing fields earn their keep on the battlefields and in the board rooms. If this defense is true, we can expect our future battlefields and board rooms to be populated only with mercenaries.

The practices of science or professional athletics are remote from the everyday experience of most people, but there are many more prosaic practices that we engage in that are also susceptible to penetration by economic considerations. Consider, for example, the games of chess and contract bridge. One or both of these games play an important part in the lives of millions of Americans. Chess and bridge are both practices. Like science, they are coherent, complex, and social, with sets of internal goods and standards of excellence that develop as the practices continue. In discussing chess as an example of a practice, MacIntyre asks us to imagine how one might introduce a young, intelligent child to the game. What reason might the child have for learning the game? The goods internal to chess—the exercise of certain highly analytic skills in imaginative and competitive fashion—cannot become clear to the child until after she has learned the game. Thus, the child's initial interest might be social—a chance to spend some time with mommy, for example, or a chance to be like mommy by doing things that mommy likes to do. If, as the child is learning, these continue to be her reasons for playing the game, then she will not become a practitioner. And her missing of the point of chess will become apparent in the way she plays. What one hopes as one initiates a child into a practice is that, as learning occurs, the goods internal to the practice begin to emerge, and that these goods replace the external ones as reasons to engage in the activity. Thus MacIntyre's description:

> . . . The child has no particular desire to learn the game. The child does however have a very strong desire for candy, and little chance of obtaining it. I therefore tell the child that if the child will play chess with me once a week I will give the child 50 cents worth of candy; moreover, I tell the child that . . . if the child wins, the child will receive an extra 50 cents worth of candy. Thus motivated, the child plays to win. Notice, however, that so long as it is the candy alone which provides the child with good reason for playing chess, the child has no reason not to cheat and every reason to cheat. . . . But, so we

may hope, there will come a time when the child will find in those goods specific to chess . . . a new set of reasons, reasons now not just for winning on a particular occasion, but for trying to excel in whatever way the game of chess demands. Now if the child cheats, he or she will be defeating not me, but himself or herself.

This description makes it clear that the distinction between internal and external goods is a crucial one. Cheating makes no sense to one in pursuit of the internal goods of chess. Lying makes no sense to one in pursuit of the internal goods of science. Both make perfect sense if the activities are viewed as means to external goods. Thus, which goods one is pursuing will have a critical influence on how one plays the game.

This has become increasingly clear in recent years in the world of tournament bridge. Bridge is a partnership game, in which partners attempt to communicate information to each other about the cards they hold. Only a very small set of communicative devices—called bids—are legal, permitted by the rules of the game. One is not allowed to communicate by changing one's tone of voice, by pausing, by coughing, by tapping the table, or by kicking one's partner underneath it. Since the potential devices for human communication are almost limitless, however, the opportunities for cheating—for conveying information illicitly—are legion. In the last decade or so, there have been major cheating scandals at the highest levels of tournament bridge almost every year. When cheating is discovered, playing conditions are changed to make that form of cheating impossible, or at least very difficult, but before long, a new set of imaginative people find ways around these obstacles, and a new cheating scandal occurs. Why all the cheating? Well, despite the fact that in this country tournament bridge is not played for money, bridge experts are able to earn substantial sums by selling their services as partners to inferior players. How much they command in fees is of course connected to their own reputation as players. Winning a major tournament can mean thousands of dollars in increased playing fees.

Which goods—internal or external—one is pursuing in playing a game will also have a critical influence on whether one keeps playing the game. External goods—candy, money, power, prestige—can always be achieved in multiple ways. At different points in history and in different cultures, different practices will provide more or less access to these external goods. And we would expect some people to drop or take up these practices as their connection to external goods waxes and wanes. But internal goods cannot be achieved in other ways. The hard core of genuine practitioners will stay at the activity through its unpredictable journey in and out of public favor. Thus, with chess, during the Bobby Fischer–Boris Spassky world championship match several years ago, Fischer's antics brought him, and the game, great public attention. Immediately, thousands of people were purchasing chess sets, going to chess clubs, and reading newspaper chess columns. But when the publicity attendant on the match subsided, so did the interest of most of these new chess players. Whatever their reasons may have been for taking up chess, they were not the reasons of committed practitioners.

The fact that people can be led into a practice by the prospect of external goods—candy, money, fame, or what have you—need pose no threat to the integrity of the practice itself. As long as those goods do not penetrate the practice at all levels, those in pursuit of external goods will eventually drop out, or be left behind, by those in pursuit of internal goods. So perhaps a handful of those who were captivated by chess when Fischer played Spassky, and a handful of those whose fathers bribe them with candy, stay with it. The rest choose to pursue their external goods in more promising domains, and the practice continues as before. However, if external goods *do* penetrate the practice at all levels, it becomes vulnerable to corruption. For practices develop, and the direction that development takes will be determined by participants in the practice. Sometimes developments are so striking that they require a fundamental reorganization of the practice—a change in the rules of the game. But what form the rule changes take will depend upon what goods participants are trying to realize.

Consider, for example, the game of basketball. The goods internal to basketball were once taken to involve adept dribbling and passing, accurate shooting, and above all, careful coordination among members of the team. It then developed that some very tall people with enough coordination to put the ball in the basket from close range started playing the game. The game turned into an exercise of muscle and height. One simply passed the ball to a tall player positioned near the basket and he shot. Now what had previously been taken to be the goods of the game became less and less important to success. To preserve these goods, the rules were changed. Players were only allowed to camp beneath the basket for brief periods of time. Most effective play still involved getting the ball to the big men near the basket, but now exquisite timing, planned movement, and sharp passing were required to do that effectively. The goods of the game had been preserved.

Contrast this rule change to another one, the introduction of a time limit within which a shot had to be taken. This rule was introduced to speed up the game, making it more appealing to spectators and thus more successful commercially. Its immediate effect was to diminish the opportunity for realizing the goods of basketball. With little time available for passing or coordinated team play, players just started launching shots from far away at their first opportunity. Eventually, skill developed so that the goods of the game were again realizable, but the point here is that the way a practice must change to maintain its pursuit of internal goods and the way it must change to maintain its pursuit of external goods need not coincide. And if the practice is controlled by people with external aims it may be corrupted to the point where it ceases to be a practice.

The distinction between internal and external goods, and the concept of a practice more generally, helps us to understand what economic imperialism means. Economic imperialism is the infusion of practice with the pursuit of external goods. This pursuit pushes a practice in directions it would not otherwise take and, in so doing, undercuts the social traditions that comprise it. It evaluates practices by a common economic denominator, abandoning the ones that fall short and encouraging the ones that do not, without regard to the internal goods that each practice possesses uniquely. And whenever internal and external goods conflict, economic imperialism moves in the direction of maximizing the latter, sometimes at the cost of eliminating the former.

This is what happened in the factory, as discussed in the last chapter. Weaving is a practice. There are goods internal to it that help to shape it, even though people do it to earn their livelihood. The pursuit of livelihood doesn't completely determine the character of weaving just because of these other, internal goods. The same can be said of furniture making, of tool making, of cooking. But when each of these practices is reduced to the assembly line, the goods internal to them disappear. Assembly-line work is not a practice. A good assembly-line worker shows up for work and works just as hard and as fast as he is told. What he is doing on the line doesn't matter. There is no developing tradition of which he is a part. The goods of assembly-line work are entirely external. This was part of the point of creating the assembly line to begin with. And economic imperialism threatens to turn other practices into the essential equivalent of assembly-line work, by moving them always in the direction of maximization of marginal productivity, utility, or profitability, at the expense of their traditional, internal goods.

For MacIntyre, the concept of a practice has a central place in a theory of what it is to be a good person. Good people possess just those characteristics, or virtues, that permit them to engage successfully in practices. The list of these virtues is fairly traditional—justice, honesty, courage, wisdom, respect, constancy, determination, and so on. And the continued existence and development of practices depends upon the continued existence of people who possess these virtues. Indeed, we could even view being a good person as a kind of "super practice"—complex, coherent, social, developing, with internal standards of excellence—that is responsible for the success of other practices. Thus, our judgment of moral worth is bound up with our judgment of the set of practices to which that worth contributes. What then happens to moral worth if the practices disappear, if economic imperialism transforms them into simply means to external goods? Well, if that happens, there is only one practice—the practice of utility maximization. And the good person will be the good utility maximizer. Virtuous, moral man will be indistinguishable from economic man. By penetrating and transforming the set of practices that comprise human, social life, economics will have created the conditions under which its conception of human nature is true.

If what it means to be moral depends in part on what our practices are, the elimination of practices will eliminate the meaning of morality. In a world made and run by economic men, moral concerns will be idle concerns. People may continue to use moral language and continue to argue about right and wrong, but the language will lose its concrete significance, and the arguments will be unresolvable. In a world like this, the battle between facts and morality discussed in Chapter 2 will be over; the language of facts will win by default. Since people will no longer possess the traditions with which to anchor their notions of good and bad, and right and wrong, the path will be open to take them to mean effective or ineffective. "Should he do this" will come to mean "is this really the way to satisfy his wants" because no other meaning will be available. The economist, behavior theorist, and sociobiologist will take prominent places as technologists, helping people to answer questions about the effects of their actions on the maximization of utility, reinforcement, or fitness. These are the people we will consult for answers to questions like "Should I tell the truth?" or "Should I divorce my wife?" or "Should I dodge the draft?" Their answers will be balance sheets, with costs on one side of the ledger and benefits on the other. This is the way moral decisions will be made.

Probably, not even the most committed economist or behavior theorist is sanguine about this vision of the world. Perhaps, they would agree, the world could do with a little more scientific advice from experts, but that doesn't mean it should dispense with moralizing. And if it is true that moral traditions depend on practices, and practices can be corrupted by the pursuit of external goods, and the pursuit of external goods is encouraged by economic imperialism, then all we have to do is be vigilant and keep economic considerations from penetrating into all our practices. By keeping practices relatively pure, we can preserve a proper place for morality in a highly industrialized, productive, and affluent culture.

Why can't we have the best of both the worlds of morality and of fact? Hirsch has already told us why. As economic activity comes increasingly to be dominated by competition for positional goods that have a social character, decisions about all aspects of life come increasingly to have an economic component. This happens whether people like it or not—whether they want it to or not. Choosing to keep the pursuit of practices pure—opting out of the race for positional goods—is not a refusal to make decisions on an economic basis. It is itself an economic choice, to forgo economic goods in order better to secure others. It is an economic choice that will make our neighbors better off and us worse off.

Furthermore, as fewer of people make that choice, and instead enter practices with external orientations, the practices themselves will change so that the pursuit of previously internal goods will no longer be possible. Return for a moment to the little experiment, in which intelligent college students treated a scientific, problem-solving situation as if it were an assembly-line job. All that the experiment really did was trick them. They knew what problem solving was, and what it required, and they knew what mechanical, repetitive activity was, and what it required, and the experiment merely induced them to treat an instance of the former as if it were the latter. It was a trick that could easily be corrected by saying: "look, you thought this was a mechanical task, but it really wasn't. It required you to think creatively and to formulate hypotheses and test them, like a scientist." We could say this, and the students would know right away what we meant.

But suppose there were no practice of science left as we know it?

Suppose science had been completely penetrated by economic imperialism? What help would it be to the students then to be told that they were supposed to be doing science? What would "doing science" mean? The economist may expect that economic imperialism can be exercised in moderation, but the problem is that there may be nothing left to provide that moderation. Karl Marx said it forcefully in this apocalyptic vision of the future:

> Finally, there came a time when everything that men had considered inalienable became an object of exchange, of traffic, and could be alienated. This is the time when the very things which till then had been communicated, but never exchanged; given, but never sold; acquired, but never bought—virtue, love, conviction, knowledge, conscience, etc.— when everything, in short, passed into commerce. It is the time of general corruption, of universal venality.

And the irony is that should this time ever come to pass, in which everything is available for exchange in the market, the market will no longer be able to function. We turn to this irony in the next chapter.

Economic Imperialism: The Market Erodes and Democracy Explodes

In any actual world there will be, for the individual cases in which he can give free reign to his personal predilictions, and others in which it will be hoped that he will draw upon his moral resources and act in accordance with ultimate ethical values rather than indulge his own preferences. . . . One of the sins committed by the glorification of economic freedom has been precisely that it has tended to confuse individuals as to where the boundary between the two cases lies.

WILLIAM S. VICKREY

The danger to liberty lies in the subordination of belief to the needs of the industrial system.

JOHN KENNETH GALBRAITH

Someone who views the growth of economic imperialism with alarm might respond to the discussion in the last chapter by saying: "It won't happen to me. Market considerations won't penetrate *my* social relations. I won't confuse private interest with public virtue. I won't confuse mortality with gain. If others want to live that way, let them. I don't live that way, I won't live that way, and the misplaced values of others won't affect me." The story to be told in this chapter is that a response like this is something of an exercise in self-deception. Economic imperialism affects everyone. It does not spare those who choose not to play the game; indeed, it extracts a price for their nonparticipation. As long as many people are playing the game, everyone is. As a result, people who think it's a bad game have to do more than refuse to play it; they have to get others to refuse to play it, too. And perhaps they have to work to reorganize social institutions so that it simply can't be played at all.

There are three reasons why we are all players in the game of economic imperialism. The first is that many of the collective bads that result from the pursuit of individual goods affect everyone. Polluted air and water make up the environment of players and nonplayers alike. So does inflation, and so do overcrowded jobs, neighborhoods, and highways.

To see the second reason, think about what it would mean to decide to resist the pull of economic imperialism. One would pursue goods that were appropriately economic in the market. The other goods— the goods of education, of family, of religion, and of civic life—presumably, one would seek by participating in the appropriate, nonmarket, social institutions. But what happens to the pursuit of these other goods if economic considerations start to pervade these nonmarket institutions and transform into nothing but extensions of the market? It makes the pursuit of noneconomic goods very difficult. Indeed, it threatens to make the very notion of a noneconomic good disappear.

One might, for example, be interested in pursuing a liberal arts education and discover that there are no institutions left that provide it. The spread of economic imperialism poses precisely this threat to noneconomic institutions. Perhaps paradoxically, one of the institutions threatened by economic imperialism is the institution of the market itself. The market cannot function in the smooth and efficient way that economists say it should if people actually behave as thoroughly economic organisms. In short, the spread of economic imperialism undercuts the very institution that has given rise to it. Not only does it make the pursuit of noneconomic goods difficult, but it even makes the pursuit of economic goods difficult. Thus, economic imperialism threatens to erode the market system.

There is another institution whose character as we know it is threatened by economic imperialism, and it represents the third reason why people can't avoid playing the economic game. It is the institution of political democracy itself. In recent years, a view of democratic politics known as the "economic theory of democracy" has become increasingly popular among political and social theorists. It treats democracy as nothing but a market in which citizens "buy" social policies with their votes. Many of the problems with our democratic institutions that have surfaced in recent years can be traced directly to the idea that the voting booth is nothing but a department store. Furthermore, if people continue to treat the democratic state as a market, democracy may explode under the pressure of millions of self-interested individuals making competing and often incompatible demands. Clearly, this is an outcome in which everyone has a stake, whether people as individuals choose to play the economic game to the hilt or not.

The threat that economic imperialism poses to democracy is a second paradox. For defenders of the market, one of its cardinal virtues is the freedom and democracy it affords to economic actors. Yet freedom and democracy in the market may undercut freedom and democracy in society as a whole.

The Tipping of Sociality and the Changing of Social Norms

Think about what it would look like if people approached their relations with others from the stance of economic rationality. What would it mean to regard social interactions as acts of "exchange"? If people were actually exchanging commodities, the tenets of economic rationality would be clear. People enter into exchanges only if they expect to benefit from them; they must be better off after the exchange than before it. The exchange must bring them closer to maximizing their self-interest. Pursuit of self-interest is, after all, the engine that drives the economic machine. Applied to relations with other people, this economic orientation prompts people to ask, "What's in it for me? How will I gain from this interaction? Is it worth 'spending' the time?" Relations with other people are a means to one's own, personal satisfaction. Just as stereos, fine dinners, cars, and the like are to be used to satisfy desires, so are people.

Does anyone really treat people as nothing more than objects of desire? Certainly, people rarely do so blatantly, but that may be because even if people are treated as objects, there is something special about them that makes them importantly different from other commodities. Unlike stereos, at the same time that people are objects, they are also *subjects,* looking at other people as potential objects of their desires. Stereos have no choice about what is done to them by whom; people do. If everyone is out in search of others who will satisfy their desires, the social relations that emerge will often be those that are mutually satisfying. If they aren't, one of the participants in the relation will terminate it. Because people are free to move in and out of social relations, they won't settle for relations in which they are pure objects, satisfying someone else but getting no satisfaction in return. It's "I'll scratch your back and you scratch mine—or else I'll leave."

The theory of natural selection provides a good model for the way these kinds of social relations might form. People careen about in society as isolated atoms, more or less at random. They bump into other people. If the collision does nothing for either of them, they keep careening. If it does something for only one of them, that person will want to maintain contact, but the other will not. Only those collisions that do something for both people will be selected—will stick. To treat another person as an object is to treat that other person as one's slave. Because people aren't allowed to do this (however much, as economic men, they might like to), much of the vulgar edge is taken off social relations, even if people do think of others strictly in terms of what they can do for them.

An alternative to regarding other people as objects of desire, as articulated by Immanuel Kant, is to take as the first principle of morality that people always be treated as ends, never as means; always as subjects, never as objects. People are to be valued and respected in and for themselves. The value of a person is independent of his usefulness to others. In essence, this is a commandment to regard other people as we regard ourselves. We should be concerned for the well-being of others, and not because of how their well-being makes *us* feel, but because as people they are entitled to that concern. This distinction echoes Sen's distinction between "sympathy" and "commitment." We should approach social relations with other people not asking, "What's in it for me?" but instead, "To what is this person entitled by virtue of being a person?"

This alternative may seem more than a little unrealistic. Only saints treat all people in this way. Even if human beings are not necessarily all greedy egoists, it is certain that they aren't all saints. Adopting this kind of moral stance toward other people is hard work. It requires eternal vigilance. It's a full-time job. Besides, unless everyone adopts this stance, those who do will be trampled by those who don't.

These concerns are all legitimate. It *is* hard work to maintain this moral stance toward other people. Indeed, without help it may be impossible for individuals consistently to act in this way. But there is help, or at least, there was help, in the form of various social institutions that function to make this moral stance almost automatic.

This point was made by Aristotle, in the *Nicomachean Ethics,* two millenia ago. Aristotle noted that it is hard to be good, and what makes it possible is the moral education people receive—in the home, in the school, in the church. This is not moral education in the abstract but in the concrete. People are taught very specifically how to act in the various situations they are likely to encounter in daily life. They are taught what it means to show respect and concern for others. Like multiplication tables, appropriate action is drilled into people until they perform it quite automatically and effortlessly. When training is complete, inappropriate action doesn't suggest itself as a possibility, just as it is not a possibility that two times four might be nine.

Of course, a moral education can't cover all the situations a person may encounter in life. Sometimes, one will have to work to be moral. But at least it makes being moral easy in the most commonplace situations, and it inculcates in people a model of what being moral means that they can turn to when necessary. One of the most serious concerns about economic imperialism is that as economic considerations penetrate social institutions that once provided moral education, the education will suffer. As a result, being moral will become harder and harder, less and less automatic. Even if people have never been trained to multiply fifty-nine by one hundred thirty-seven, the multiplication training they have received makes this a manageable task. In the absence of that training, the task would be impossible.

In addition to seeming unrealistic, the Kantian alternative may also seem overly romantic. Are people to show no concern for self-interest when they interact with others? Should people participate in social relations without any regard to what they get out of them? Surely it is appropriate to want to get some satisfaction out of relations with others. People should expect to benefit in some way from their social activities. But if that's true, why not be economic about it and calculate the potential costs and benefits of this or that social activity?

The reason why not is that the source, the nature, and the temporal character of the benefits people derive from social activity are often not amenable to economic calculation. In economic activity, people seek and expect fair and profitable exchange in each transaction. When someone buys a stereo, she wants her money's worth from *that* stereo. It is small comfort to know that a purchase contributes to the well-being of the market system in general, making possible many future profitable exchanges, if this particular exchange is unsatisfactory. Economic calculation is focused on this exchange, with this person, at this time.

The situation is often quite different in the social domain. People do not analyze a marriage into a series of exchanges and ask whether each social action of theirs has been appropriately compensated. They do not have quid pro quo expectations of spouses. What matters is some assessment of satisfaction with the marriage as a whole. There is no reason to believe that this long-term, overall assessment of a marriage is just the sum of a very long and large series of particular exchanges—sex in return for doing the dishes on Tuesday, a special dessert in return for doing the marketing on Thursday, and the like.

Furthermore, some of the social benefits people derive come from relatively anonymous others: the collective enthusiasm of the home-team fans at a sporting event; the tranquility of a house of worship; the heated interchange of ideas in a college seminar. None of these goods represents payment for particular social actions. If they are payment at all, the payment is for acting socially in general. The more people act like Kantian social beings, the more they benefit from the social institutions in which they participate. So while we may not care that our stereo purchase helps support the market system, we certainly do care that our seminar participation helps support the educational system.

Thus, many of the social benefits people obtain derive less from the particular others with whom they interact than they do from the character of the social institutions that frame the interactions. And the danger of economic imperialism is that it will change the character of these social institutions so that the benefits people derive from them diminish. As more and more people approach social activities economically, the norms that govern social activity will change to reflect this economic approach. As the norms change, the institutions will change. As the institutions change, both the benefits and the moral education people derive from them will change. The end result will be that it costs more and more to be noneconomic in social interactions. As Fred Hirsch observed: "To act socially is less costly in a social setting. . . . Differently put, there is an interdependence in social orientation itself. Its costs to the individual are indeterminate without knowledge of how other individuals act."

This point can be made somewhat clearer by reference to particular examples. Consider sex. What kind of a good is sex, and what sort of stance should people adopt toward it? Clearly, it can be viewed as a strictly economic commodity. People can purchase the services of prostitutes just as they purchase the services of plumbers. When viewed in this way, sexual activity is discrete and impersonal. If there are goods to be gotten from an evening with a prostitute, they will be gotten there and then. Sure, they may linger for a while, the way the taste of a good wine lingers, but they will not have pronounced or permanent effects on future relations with this or any other prostitute. And it matters very little, to either the customer or the prostitute, who the sexual "other" is. Both buyer and seller are interchangeable with other buyers and sellers. People are not indifferent to whose services they are buying. Some prostitutes are more skilled than others; some are more attractive. But it is just a collection of relevant attributes, and not a person, whose services are being bought.

On the other hand, sex can be regarded as an act of spiritual as well as physical union, as an opportunity to drop barriers, show vulnerability, and establish intimacy. Viewed in this way, it matters a great deal who one's sexual partner is. People are not interested in being this close, and this exposed, with just anyone. Furthermore, episodes of sexual union contribute to the building of deep and lasting personal relations. There are cumulative effects that make it important that sexual activity be with the same partner. And these cumulative effects make it possible to reap social "goods" from sexual activity far into the future. Good prostitutes are skilled at *simulating* the care and concern for their clients as individuals that are a part of noncommercial sexual relations, or even of showing genuine concern—that ends when the night is over. In this respect, they are much like psychotherapists. A hundred dollars may buy an hour of sympathy and understanding. But no matter how genuine the concern may be, it is money, not concern, that binds the relation, a fact that is sometimes made painfully plain to people in psychotherapy when they are told by their therapist, while deep in the midst of an especially troubling revelation, that the hour is up. If either the client or the prostitute (or therapist) oversteps the implicit contractual bounds of the relation, there is hell to pay. This is vividly demonstrated in the movie *Klute,* when the prostitute who is the main character finds herself in need of assistance from one of her regular clients. He can offer her neither himself, nor his time, nor his emotional support. All he can offer her is some money.

Now prostitution is illegal almost everywhere in the United States, so it seems clear which view of sexual activity is the "official" one in our society. But some militant feminists argue that marriage is just legal prostitution in disguise—sexual services (and cooking and ironing) in exchange for food, shelter, and protection. Is it? Certainly,

sexual relations within marriages can be more or less commercial—
more or less spiritual. People differ both in what they expect from
their marriages and in what they get from them. This wide range in
the nature of legal and socially sanctioned sexual relations might lead
one to the opinion that sex should be neither commercial or spiritual.
It should be what the people engaging in it want it to be.

This is the standard, liberal view of sexual activity, but it is mistaken. What an individual does in the bedroom does affect others,
and not just the others he does it with. It affects others because
individual sexual activity contributes to the development of social norms
about what can be expected from sexual activity in general. If the
belief that marriage is just a devious route to commercial sex becomes
widespread, then individuals will have no choice but to view it that
way themselves. Someone who holds out and seeks intimacy, deep
emotion, and genuine concern from sexual partners will be regarded
as an odd, anachronistic romantic. So other people's views about what
sex should be *do* tread on ours; their practices affect our aspirations,
and their aspirations affect our practices. If enough people believe
that marriage is just legalized prostitution, it will become legalized
prostitution, even if it wasn't initially. The status of sexual activity
as a good is something that must be worked out—struggled over and
debated—by society as a whole. It simply can't be all things to all
people.

Now consider a second example, one discussed by Michael Walzer
in his book, *Spheres of Justice*. It concerns the evolution of an institution that has become commonplace in modern, industrial society.
The institution is known as the "vacation." The "vacation" is a modern phenomenon, perhaps a century old, and its extension to any but
the very rich is more modern still. The institution that it has gradually replaced is the "holiday" (holy day), and there is perhaps no better way to exemplify the change from public, social, and noneconomic
to private, individual, and economic than in the shift from holidays
to vacations.

Dating at least from ancient Rome, the holiday was a time of public
and communal celebration, a time to commemorate some event of
civic or religious significance that all citizens participated in equally.
The set of holidays observed by a given community was a way of
defining that community. Each holiday, with its unique history and
set of rituals, connected the members of a community to one another,
and to the community's collective past. The holiday was fundamentally noneconomic in character. Everyone participated, indepen-

dently of economic circumstances. Indeed, often, everyone was *required* to participate, as in the case of the Jewish sabbath, so that the holiday did not even involve much cost in material opportunities forgone. The importance of the holiday to the structure and integrity of the community is evidenced by the fact that almost invariably, when revolutionary movements come to power, they attempt to destroy traditional holidays. In doing this, they attempt to break down sources of allegiance and loyalty that are tied to the old social order and represent a challenge to revolutionary authority.

In contrast, the vacation is thoroughly private and economic. People negotiate for paid vacations with their employers. They decide whether to spend their money on vacations or on things, in a way that they never would with holidays. Imagine asking whether to buy a new car or celebrate Easter. The point of a vacation is not to join in celebration with other members of the community but to escape it—at least for a while. People take vacations for a change of scene, and they take them alone, or just with their families. When vacations are at different times, and when they take people to different places, they do not contribute, as holidays did, to keeping the community together.

As must be obvious, holidays are disappearing from American life. Within the rhythm of the week, there is no longer a period when everyone rests, or has the free time to join with others in some kind of communal activity. Shops are open seven days a week, and "weekends" occur on different days for different people. The few national holidays we do observe are increasingly being separated from their histories, from the dates they are intended to commemorate. The birthdays of George Washington and Abraham Lincoln are now observed on a "Presidents' Birthday" weekend, always a weekend so that the number of people who can actually escape is maximized. If they signify anything, it is not the lives of great figures in American history but the opportunity to go on ski trips and to take advantage of bargain prices in retails stores.

Certainly, the vacation offers advantages that the holiday does not. The holiday does not offer freedom from having to be at a certain place at a certain time doing a certain thing, as is the case on the job. People may be at a different place, doing a different thing, but their obligation to participate in the holiday is no less strong than their obligation to participate on the job. The vacation, in contrast, truly does afford a kind of liberation. However, the benefit of the vacation carries with it an attendant, somewhat hidden cost. It eliminates a

major source of communal identity and strength. And as was true of
sex, one man's vacation cannot be another man's holiday. If a significant component of what makes something a holiday is communal
participation, the failure of some people to participate has an inevitable effect on others. As people choose to turn holidays into vacations, our own holidays become vacations whether we want them to
or not. Communities can of course decide to take their leisure as
some mix of holidays and vacations, but for holidays themselves to
persist, there must be community agreement on what they are.

Economic imperialism contributes to turning sex into a commodity
and to turning the holiday into a vacation. The general consequence
of developments like these is the undermining of public life and of
the social institutions that support it. When an individual decides to
treat sex and free time as private, economic domains, the decision
affects not only his behavior but the behavior of others who have not
made that decision. The social norms governing these activities slowly
change, making it increasingly difficult to support a public, social,
and noneconomic stance toward them. Eventually, one has no choice
but to "consume" sex and vacation time. Without the social institutions that depend on public virtue, on commitment and concern for
others, there is no reason for such an orientation to the general good
to persist. There is no opportunity for the development of "automatic
goodness." Eventually, it must disappear all together. When this
happens, people have no choice but to turn to the competitive marketplace, as a supplier not just of food, clothing, and other commodities but as a supplier of goods that used to be provided in other ways.

But when people turn to the marketplace, they are in for a shock.
It can't do a very good job of supplying these previously noneconomic
goods. Indeed, once these previously noneconomic goods becomes
economic, the marketplace can no longer even supply people very well
with run-of-the-mill commodities. For the marketplace itself as an
institution depends upon the existence, within its participants, of
some measure of public virtue—of commitment to doing what is *right*,
and not just what is profitable. The very same market system that
has given energy and direction to economic imperialism is threatened
with destruction by it. As our public, social, noneconomic orientation
erodes, the market erodes with it.

The Erosion of the Market

The autonomous, free, self-regulating market is a myth—an economist's abstraction. Such a market has never existed in the past and does not exist in the present. Governments have always had a hand in regulating the market. As Fred Hirsch observed:

> Market capitalism has never been the exclusive basis of the political economy of any country at any time. That has been its strength. It was the marriage of market capitalism with state regulation that produced a hybrid politico-economic system with the necessary resilience and plasticity to survive.

In Western history, the mix of government regulation with individual enterprise goes at least as far back as ancient Athens. There was plenty of commerce in Athens, but the state interceded to assure that everyone was provided with the necessities of life. With respect to food, there were commissioners of trade, inspectors of weights and measures, regulators of prices, and when times were hard, officials who oversaw price reductions and rationing. There was also regulation of public health, of funerals, of roads. And the wealthy were taxed so that all citizens could attend the great drama festivals. In ancient Greece, it seems, the theater was considered a necessity.

Just prior to the industrial revolution, there was regulation of the three essential ingredients of a market economy—land, labor and money. Usury was either tightly controlled or prohibited. Land was often not allowed for sale, and if allowed, only to select members of the community. And labor—what one did, how one did it, and what one was paid—was highly circumscribed by a network of social customs and legal obligations.

In general, matters of public regulation and provision have centered on two different questions. The first question concerns what should be regulated and provided. What is to count as a genuine need? The second question concerns how much can fairly be said to satisfy that need. Neither of these questions has a straightforward or simple answer. Indeed, the way societies answer these questions goes a long way toward defining them. It reflects the values of a society, the practices that bind its members together, the moral code by which it is governed. If societies feel bound to provide for the genuine and essential needs of their citizens, the particular needs they regard as genuine and essential tell us what it means, in that society, to be a person. In Athens, a person could not be deprived of the theater.

Walzer puts it this way, in describing the nature of the social contract that each society develops to provide for its members:

> It is an agreement to redistribute the resources of the members in accordance with some shared understanding of their needs. . . . The contract is a moral bond. It connects the strong and the weak, the lucky and the unlucky, the rich and the poor, creating a union that transcends all differences of interest, drawing its strength from history, culture, religion, language, and so on. Arguments about communal provision are, at the deepest level, interpretations of that union.

The ideology of the free market insists that if regulation is replaced with the pursuit of rational self-interest, the result will be more efficient, more productive, and more fair. Now, of course, the modern American economy is far from a free market. There is government regulation and control everywhere, in part to correct for market imperfections and in part to meet perceived responsibilities to the citizenry. And there are other, nongovernmental controls on market activity that are no less important, if somewhat more subtle. These are noneconomic moral and social controls that stem from social norms governing appropriate conduct. When Adam Smith was writing, these social norms were firmly established and easy to recognize. Thus, he could be sanguine about the "natural sympathy" that would lubricate and cushion market activity. His mistake was that this sympathy is not "natural"; it is a social product whose very existence is being undermined by economic imperialism.

What are some of the moral, social supports on which the market depends? It depends on an implicit agreement about what can be bought or sold. It depends on the assumption that people tell the truth and that they honor their contractual obligations. Imagine what the marketplace would be like if people had to do research before each purchase because people couldn't trust what was on the label. Imagine what hiring employees would be like if people had to check out each and every item on the applicants' resumes because they couldn't be counted on to tell the truth. Each of these moral and social agreements is buttressed by law, but we have already seen how laws of this kind are completely unenforceable once they need to be enforced for everyone. Enforcement of the law works when only a handful break it. As it becomes increasingly routine for people to misrepresent themselves or their products, or to violate contracts, the legal system will be overwhelmed as it attempts to catch and punish transgressors.

Even if we could afford the massive legal system that would be required to police a society full of completely self-interested individuals, how would we control the behavior of the enforcers? How could we stop judges and juries from becoming commodities? If judges started selling their services to the highest bidder, the entire system of private property, on which market activity depends, would collapse. There would be no means of determining "rightful ownership," except, perhaps, by force. As Kenneth Arrow puts it, "The definition of property rights based on the price system depends precisely on the absence of universality of private property and of the price system." It appears that the market system really can't do without the system of moral and social constraint that Adam Smith took for granted.

But perhaps we don't have to appeal to morality. Instead, we can appeal to self-interest—*enlightened* self-interest, perhaps, but self-interest nonetheless. If people know how much their dishonest, unrestrained pursuit of self-interest is costing society, and them, as taxpaying members of it, they will mend their ways. They will see that it makes good *economic* sense to tell the truth, honor contracts, and the like. When they learn that malpractice suits have increased the cost of their visits to the doctor by 10 or 20 percent, they will think before they sue. When they see how much it costs to have a legal staff on retainer as a hedge against breach of contract or product liability suits, they will honor contracts and produce safe products. When they see that income tax dodges simply raise everyone's tax rates, they will stop seeking and using these dodges. Won't they?

The answer is no, they won't. For the general problem with voluntary, collective action—the free-rider problem—will rear its ugly head. The actions of an isolated individual are neither necessary nor sufficient to affect the practices of society as a whole. One person's willingness not to sue for malpractice won't change the cost of malpractice premiums. One person's commitment to honoring contracts and making safe products won't change the practices of others. One person's willingness to refrain from using tax dodges won't change the tax rate. As long as a single individual alone shows "enlightened" self-interest, he will be much worse off than before. For it to be effective, everyone has to do it. But once everyone, or almost everyone, is doing it, individuals can really make out by continuing to sue for malpractice, to use tax dodges, and to make unsafe products. As long as self-interest is the sole motive for action, the most effective strategy is to take merciless advantage of the restraint or public commitment shown by others.

It was the implicit recognition that a collection of individuals running around in pursuit of their interests will not ensure the common good that led John Maynard Keynes to his revolutionary "state-managed capitalism" fifty years ago. Some government tinkering—to stimulate demand and thus spur economic productivity—would be necessary. But Keynes underestimated the need for government control because he assumed that, unlike the men on the street, the captains of industry had motives considerably loftier than self-interest. Influenced, no doubt, by the fact that so many of these "captains" in England came from the British aristocracy, for whom, as a class, the pursuit of profit was a bit crass, Keynes assumed that the leaders of the economy would themselves show restraint in seeking profit for their companies. They would consider and act in the public interest. Not that their business activity was a charity, by no means. But there were bounds on what a business could legitimately do in the service of profit, bounds dictated by custom, and by an aristocratic tradition of public service.

Here, Keynes was guilty of the same mistake made by Adam Smith. He assumed that the social and moral fabric of his society was a part of the natural order of things. He failed to realize that the very act of participation in the market system would turn his lofty aristocrats— or if not them, then their children—into self-seeking egoists no different in motivation from the people they employed. Certainly, the era of "captain of industry as public citizen" is on the wane at present in the United States. A recent article in the financial section of the New York *Times* (August 19, 1984) decried the modern era of "me-first management"—of hostile takeover, of corporate blackmail, of high rates of executive turnover, and of the pursuit of enormous short-term profit without any regard for the long-term well-being of either the company or the society as a whole. The dominant attitude of high-level management at present seems to be "get as much as you can as fast as you can, and let someone else worry about picking up the pieces." But who is that someone else, and where will he come from? The free rider has come to roost at the pinnacle of the economy.

Furthermore, long-term well-being is not the only thing the captain of industry ignores in pursuit of short-term profit. With growing frequency, he also ignores the law. As discussed in another recent New York *Times* article (June 9, 1985), white collar crime is rampant, and it is occurring at the highest managerial levels of major corporations. Many recent and dramatic examples of high-level white collar crime are documented by Mark Green and John F. Berry in their book, *The Challenge of Hidden Profits: Reducing Corporate Bureaucracy and Waste*. They include the defrauding of the government by several large defense contractors, check kiting by a major Wall Street brokerage house, drug money laundering by a major bank, tax fraud by a jewelry house of international fame, price fixing by a major blue jean manufacturer, bribery of foreign government leaders by aerospace firms, violation of waste disposal laws by chemical manufacturers, and production of products that are known to be unsafe by automobile manufacturers. The list could go on; it is distressingly long. Clearly, the captains of these industries are not the kinds of people that Keynes had in mind.

If the pursuit of profit fails to foster civic-mindedness among industrial leaders, what does, or might? One of the strongest traditional sources of restraint and moral strength has, of course, been religion. Religious institutions have traditionally placed great value on acts of other-regarding altruism. Religious teaching has helped establish the "automatic goodness" that influences people to internalize a commitment to honesty and a concern for others so that not all rules need external enforcement. Paradoxically, it has been the pull of religion, the concern for how one will make out in the next world, that has contributed mightily to material well-being in this one. Hirsch puts it this way:

> The payoff to religious belief is in earthly coin. The traditional concept of religion as insurance on the next world, which might or might not pay off in this one, is exactly reversed. One might or might not go to heaven by loving one's neighbor as oneself. . . . What was certain was that one would thereby get more worldly goods out of the market; provided that all one's neighbors did likewise.

The problem is that economic imperialism undercuts these very religious institutions that give aid and comfort to the market. It makes altruism and honesty a game for suckers. The point is apparently lost on modern members of the "moral majority," who simultaneously champion the free-market economic system and the return to traditional moral values, failing to realize that the former undermines the latter.

There have always been some economists who knew that the free market, divorced from noneconomic influences on action—divorced from personal morality—could not survive. In the words of Lionel Robbins, an influential economist from earlier in the century, "A theory of economic policy, in the sense of a body of precepts for action, must take its ultimate criterion from outside economics." And the welfare economist A. C. Pigou noted that the market system depends upon "a stable general culture, [in which] the things outside the economic sphere either remain constant, or, at least, do not vary beyond certain limits." Economists have known this, but they have not worried about it. They have worried about the autonomous economy and assumed that noneconomic sources of influence and restraint would take care of themselves. But they won't take care of themselves. Instead, they are being taken care of by economic imperialism. It is increasingly hard to find criteria for economic policy that are "outside economics" as more and more social institutions are brought inside economics. And it is increasingly mistaken to assume that things outside the economic sphere "remain constant" as they are continually being invaded by things inside the economic sphere. Just when we need them most, these noneconomic sources of restraint and stability of the market system are disappearing. Thus it is that economic imperialism erodes—or corrodes—the common core of moral conduct on which the market depends.

Economic Imperialism and Democracy

There are two quite distinct lines of argument available to defenders of the market system. One might be called "scientific," and the other might be called "moral." The scientific defense is that the free market is the most efficient, effective, and productive system known for allocating resources; it is Pareto optimal. We have devoted a good deal of space considering the pros and cons of this scientific line of argument. The moral defense is that the free market, and only the free market, is consistent with the principles of freedom that citizens in a democracy hold dear. No one can be told where to live. No one can be told what to wear or to eat. No one can be told what work to do. No one can simply take what rightfully belongs to someone else. All exchanges—of things or of labor time—are free and voluntary. Any encroachment on the market system—any externally imposed restraint on market activity—is a restraint on freedom of choice. Thus, a market economy is the only system consistent with a democratic polity.

This marriage of a free-market economic system and a democratic political one seems natural enough. Democratic freedom would certainly be incomplete if people were free to say what they wanted, read what they wanted, and worship what they wanted, but not to buy what they wanted, or charge what they wanted for what they were selling. Indeed, the marriage seems so natural that in recent years people have come to use the metaphor of the marketplace to help explain the nature of democratic political organization. According to the resulting "economic theory of democracy," citizens "shop around" in the political marketplace, assessing the various commodities (social programs) that are available. They ultimately "buy" (vote for) those commodities that maximize their individual preferences. Given this naturalness, it is more than a little ironic that this economic approach to democracy now threatens to undermine the democratic political system. American democratic institutions are on the verge of exploding, in part because people have come to treat the state as just another free market. For the effective functioning of a democratic state depends upon more than just the pursuit of individual interests by its citizens. It depends on people's willingness to behave like noneconomic men.

Historically, the roots of American democracy were grounded in morality. Discussions of "justice," "equality," "freedom," and "authority" derived from ideas about right and wrong and presupposed public commitment and responsibility. As citizens, people owed it to the state to be concerned for the public interest. People owed it to the state to be loyal. People owed it to the state that they would participate in public life, at least by voting. The right to vote was more than just a right; it was a responsibility. The franchise was *inalienable;* people couldn't sell their votes for cash even if they wanted to. Participation in a democracy was thus a kind of moral calling, and a democratic state would not be worthy of the name unless the bulk of its citizens honored that calling. The state truly served the people only if the people were willing to serve the state, by accepting the responsibilities that citizenship demanded. The preservation of democracy itself depended on this moral core, and the resolution of disputes about the extent of government responsibility depended on a common moral vocabulary. Most accounts of democracy have argued that democratic political systems cannot be sustained in the absence of a set of shared values and moral precepts that unite the citizenry into a community. Democracy, it seems, is another example of what MacIntyre calls a practice.

At least, that's the way things were historically. They seem to be no longer. Instead of being grounded in principles of morality, democracy is coming to be grounded in principles of the market. That is what the economic theory of democracy is about. Let us, then examine what a state run exclusively on market principles, with no pre-sumption that people are guided by anything but self-interest, looks like, and what its shortcomings are.

The origins of the economic theory of democracy can be traced to economist Joseph Schumpeter, in his book *Capitalism, Socialism, and Democracy*. According to the theory, the political sphere of life is accurately modeled on the economic sphere, with both spheres inhab-ited by the same self-interested, preference-maximizing, economic men. Groups of political entrepreneurs—the elites—seek the rewards of political power by offering goods (programs) in the political market. Larger groups of people—the masses—purchase (vote for) the goods they want. Not only *can* people sell their votes, but they do so rou-tinely, not, of course, for cash, but for programs that are eventually convertible into cash. Inalienability of the vote now means that peo-ple can't sell their votes to just anyone. Only elites (candidates for office) are potential purchasers. Political participation is essentially the making of demands on government, and the democratic process is a way of aggregating individual political demand in the same way that the market aggregates individual economic demand. And the aim of the elite is to stay elite, that is, to stay in power. To this end, they offer whatever goods it takes to "buy" the votes of the masses.

This very simple theory of democracy has the virtue that it has very low expectations and makes almost no demands of the citizenry. Morality, public commitment, and even serious political participation are unnecessary. All that is required is the pursuit of self-interest, which is presumed to be what human nature is about anyway. How-ever, this simplicity has a price; it doesn't work. The analogy between government and the market is not perfect.

The market, when it is functioning smoothly, is governed by an invisible hand that ensures that supply will equal demand. The price mechanism sees to this equilibration, with prices rising or falling as demand and supply rise and fall. The government has no such built-in equilibrator. There is nothing to check the growth of aggregate demand. Thus, individual interest groups or lobbies keep yelling "more, more, more," and "me, me, me," and the elite will promise anything to stay in power. Superfluous military bases should be closed, but not the one in my congressional district. Extravagant defense contracts should be eliminated, but not the one that keeps the local company in business and employs five hundred voters. Social programs should be cut. Which ones? Not support for the elderly, says the senator from Florida. Not support for minorities, says the senator from New

York. Not support for farmers, says the senator from Iowa.

Once, these various demands could be kept in check by the mechanism of party politics. The leadership of each party could influence party members to toe the line and submerge their own special interests for the good of the party, if not the country. Then, political debate and horse trading had only two voices—the Democrats and the Republicans. Now, individual legislators are much less party loyalists and much more free agents. Congressional debate entails five hundred voices, all screaming at once, often for incompatible things. Little gets done, and the country simply can't afford to satisfy everyone's demands. In the political market, demand now vastly exceeds supply. The result is what is variously called an "overload of democracy" or a "crisis of democracy." America can no longer afford all the democracy it has. We can't keep buying everyone's votes. Excess demand must somehow be checked.

But how can it be checked? If a member of the elite says no to enough demands, he will not be elected to carry out his policies. People will elect a "yes man" instead. And if the "yes man" then goes back on his promise, he won't be reelected. It seems that as long as people are selling their votes for programs, there is nothing to check the impulse not to sell cheap.

In a recent article on the current crisis of democracy, Douglas Bennett and Kenneth Sharpe identified some potential solutions to this problem that have been suggested by others. First, government can curb severely, and across the board, its role in meeting human needs. This is the strategy being pursued by the Reagan administration. Second, government can take some democracy away—replace electoral institutions with some more authoritarian ones that can do what has to be done without worrying about popular support. This is obviously a solution that no one likes to contemplate, which leads to a third possibility; government can try to get individual people, or interest groups, to restrain their demand voluntarily for the greater good of everyone. Thus, as a proponent of the economic theory of democracy, Samuel Huntington, has said, "The effective operation of a democratic political system requires some measure of apathy and noninvolvement on the part of some individuals or groups." This is a solution that everyone finds more palatable. However, as Bennett and Sharpe argue, it is a solution that is radically incompatible with the economic theory of democracy itself. Not only is there no basis within the theory on which to call for voluntary restraint, but even if we called for it, we wouldn't get it.

Their argument has much in common with the views of Fred Hirsch already discussed. First, they say, there are two possible grounds on which to appeal to people to restrain their demands. One can appeal to their self-interest, or one appeal to a set of democratic values about the common good that they all share. Appeals to self-interest won't work because of the free-rider problem. Even if people were convinced that social cooperation and restraint were essential to preserving democracy, and that democracy was the best possible political system for the pursuit of self-interest, it would be in the best interests of every individual to be a free rider, to let everyone else exercise restraint.

This problem of the free rider has come up repeatedly. It may be immoral to let other people shoulder the burden, but it is certainly not irrational—as long as people can get away with it. The logic of the free ride applies to paying taxes, honoring contracts, and restraining demand on government. As long as people view themselves as isolated, economic agents, with the maximization of self-interest as the only legitimate motive for action, the free-rider problem will only grow. The more imbued people are with the economic theory of democracy, the less responsive they will be to appeals for voluntary restraint of demand. Just as the unbridled pursuit of self-interest will cause the market economic system to erode, it will cause the democratic political system to explode. And this will happen although all participants are behaving as rationally as they can. Indeed, it will happen *because* people are behaving as rationally as they can. The lesson, again, is that both the market and the state depend for their continued existence on the exercise of economic *nonrationality,* or better of noneconomic rationality.

If appeals to self-interest won't secure restraint, what about appeals to shared democratic values? Can appeals to trust, compassion, social responsibility, and public-mindedness get people to restrain their demands? This also won't work, for reasons identified in our discussion of the corruption of practices and the spread of economic imperialism. Where are these shared values to come from? That they don't come from the economy is clear. But they also aren't coming any more from the family, the school, or other social institutions that have grown increasingly commercialized. Indeed, one place they might have come from is the political sphere itself. But it is just the commericalization of this sphere that has produced the problem in the first place. The only value people may be said to share in a world of commercialized social institutions is the value of individual preference maximization. And this is a value that produces excess demand rather than restraining it.

So it seems that voluntary restraint is not a possible solution to the crisis of democracy, at least not as long as democracy is conceived and practiced in accord with the market model. The alternatives seem to be somehow to abandon the market model or to abandon democracy. The second is unattractive, but the first may be unattainable.

There is an extraordinary irony in this turn that democracy has taken in its commercialization. Economic historian Albert Hirschman, in the book *The Passions and the Interests,* discussed how the notion of "interest" as distinct from "passion" was born. Some restraint was needed on the monarch's reckless attempts to satisfy his passions that led him and his subjects into one foolish venture after another. Some order and stability was needed in life. That stability came, it was thought, by substituting interest—rational, commercial interest—for passion. The pursuit of rational, commercial interest would be orderly and predictable rather than capricious. It would be benign rather than malignant. The clockwork of the market was a remarkably civilized alternative to what it was intended to replace. But now we find ourselves in need of something to check, restrain, rationalize, and civilize the pursuit of self-interest. It turns out that the pursuit of self-interest has much in common with heroin. Both are human inventions. Both slowly destroy the fabric of whatever social institutions they invade. Both are addictive. And heroin was introduced as a cure for the evils of morphine.

Spheres of Social Life

The dangers of economic imperialism are very real. In the long run, neither the market itself nor the democratic political system we inhabit will be able to survive the erosion of morality and social concern that economic imperialism brings in its wake. Is there something to be done about this? Can economic imperialism be stopped, or even reversed? The beginnings of an answer have been sketched by Michael Walzer, in *Spheres of Justice.* Walzer rejects the quest for universal principles of justice that apply across the board, to all people in all situations in all cultures for all times. What he offers in its place is the idea that what is just and fair is the product of particular cultures at particular times. Moreover, within a given culture at a given time, what is just will vary from one sphere or domain of life to another. The project of determining what justice is, then, is a project of dividing social life into spheres and then figuring out the rules of justice that should operate in each of those spheres. Justice will be particular, not general; historical, not eternal; concrete, not abstract. This notion that there are distinct spheres of social life that operate, or at least should operate, according to different rules may be essential to stemming the tide of economic imperialism.

Central to Walzer's thesis is the idea that different goods are appropriate to different spheres of social life. There is no common denominator that unites them. The goods of religious or civic participation, say, are incommensurable with the goods of material consumption. Money can't buy ecclesiastical office, and piety does no good in the market. There is no sensible way to equate so many mornings in church with so many evenings of the theater. In short, neither money, nor anything else, can buy everything.

The idea that different goods obtain in different spheres of social life is the key to why those spheres must be kept distinct. To combine them is to homogenize the goods they offer, which is precisely what economic imperialism does. Walzer's conception of different goods as appropriate to different spheres of life has much in common with the notion of a practice. Practices, recall, are complex social activities that are partly defined by the goods that are peculiar to them. These goods change as practices develop, but always they remain specific to the particular practice. Practices become corrupt when goods that are external to them get imported and start to change the nature of the activities of practitioners. So, Walzer might say, spheres of social life become corrupt when they import external goods.

Many characteristics of daily life implicitly recognize the fact that spheres of life have distinct goods. Either by social convention, or by law, or both, people are prevented from buying and selling many of these goods. Among the things people can't buy or sell, Walzer identifies the following: human beings, including ourselves; political power; criminal justice; civil rights; marriage and procreation; exemption from military service; political office; prizes and honors; divine grace; friendship; and love.

But the point of these last two chapters has been that these prohibitions on buying and selling are much less secure than we might think. Considering the cost of legal services, criminal justice is to some extent for sale. According to the economic theory of democracy, so, effectively, are political power and political office. To the more enthused among economists, and the more militant among feminists, so are love and marriage. According to Hirsch, friendship must now be purchased at the cost of opportunities to consume. With the development of artificial insemination techniques and "rent-a-womb" practices, procreation is on the verge of becoming an exchangeable commodity. And members of the wealthy classes have always managed to find (not quite illegal) ways to buy their way out of military service, or if not that, to buy their way into military service away from the front lines. The buying and selling of these goods that are "not for sale" is not yet completely open and unabashed, but there is no reason to believe that, as people become accustomed to the more subtle forms of exchange that now occur, exchange won't become more and more brazen.

And this would be a disaster. It is not that there is anything wrong with the buying and selling of goods on the market per se. However, people must be sure that the market extends only to goods that can appropriately be bought and sold—to commodities. To determine the appropriate boundaries on the sphere of market activity, people must be able to recognize what makes something a commodity. What is it that market goods share with each other that they don't share with nonmarket goods?

This is a difficult question, in part because the goods that are available on the market continue to change, but we can begin to sketch an answer. By and large, the goods that are appropriately traded on the market share these characteristics:

1. *The people who produce them are different from the people who enjoy them, except incidentally.* That the person working in a refrigerator factory owns a refrigerator is not central to the fact that he makes them. But that the people who enjoy the intimacy and emotional bonds of family life are members of the family is no accident.

2. *Producers and consumers of commodities are typically strangers.* The market relation requires no personal relations among participants.

3. *Commodities are exclusive goods, enjoyed only by their owners, or by others, only at the discretion of their owners.* Now this is not always strictly true. We can't prevent neighbors from deriving some aesthetic satisfaction from the new paint job on our house. Sometimes, goods have spillover effects on people who don't own them. Economists call these effects "externalities," precisely to acknowledge that they are external to, outside of, the functioning of exchange in the market.

4. *Because commodities are exclusive, they cannot be given away without their owners losing enjoyment of them.* The same is obviously not

true of noncommodity goods like love and friendship.

5. *Commodities are exchangeable.* Other goods, like talent or piety, are not.

6. *Commodities are produced and exchanged for the sake of profit.* Other goods, like children, education, and love, are not.

Among commodities, there is a common denominator—money or price—and there is no reason why there shouldn't be. The problem comes when it becomes possible to buy things with money that are not commodities. This is what breaks down the barriers between different spheres of social life. And the result is not just the corruption of these different spheres, but a kind of tyranny.

There is no way to assure the equal distribution of commodities among members of society without being unfair to those who have been lucky, talented, or daring enough to do better than others. However, this unequal distribution of commodities is not, in itself, unjust. It becomes unjust when money comes to buy things that are not commodities, or said another way, when everything becomes a commodity. Then, money becomes what Walzer calls a "dominant good," one whose influence extends beyond the market to all spheres of social life. Then, success in the market means success at everything. Then, success in the market multiplies into all aspects of life. Then, those who dominate the market dominate the lives of other people. If money buys not only cars and houses but education, political power, love, friendship, divine grace, respect, and prestige, then those who have money will necessarily tyrannize those who do not. Pascal made this observation more than three hundred years ago:

> The nature of tyranny is to desire power over the whole world and outside its own sphere.
>
> There are different companies—the strong, the handsome, the intelligent, the devout—and each man reigns in his own, not elsewhere. But sometimes they meet, and the strong and the handsome fight for mastery—foolishly, for their mastery is of different kinds. They misunderstand one another, and make the mistake of each aiming at universal dominion. Nothing can win this, not even strength, for it is powerless in the kingdom of the wise. . . .
>
> *Tyranny.* The following statements, therefore, are false and tyrannical: "Because I am handsome, so I should command respect." "I am strong, therefore men should love me. . . ."
>
> Tyranny is the wish to obtain by one means what can only be had by another.

An extreme example of tyranny by a dominant good is provided by the town of Pullman, Illinois. Pullman was a company town, and the company was George Pullman's railroad car company. There are plenty of company towns around, towns most of whose residents are employed by a single company. In towns like this, company and town—market and polity—are inextricably linked, since what is good for the company is good for virtually all of the town's residents. But Pullman, Illinois, was a different kind of company town. Not only did everyone there work for the company, but the company literally owned the town. Pullman built the town from scratch, creating houses, shops, medical facilities, schools, playgrounds, a hotel, a theater, a library, and a church. And he owned it all. The town was governed in the same way a stereo is governed—by its owner, as a property right. Pullman had clear ideas about how people should live, and he imposed his ideas on the residents of the town. A certain measure of tidiness and decorum was expected, and violators could be fined by Pullman's inspectors, or worse, evicted on ten days' notice. Alteration of houses was strictly controlled, and no private construction was permitted.

Pullman's intentions were benevolent, and he ran the town with a generous, if iron hand. The housing was good and was kept in good repair. The rents and prices were modest. But benevolent or not, all decision and control rested with one man—not the man chosen by the people, but the man who owned all the resources. That this mode of governance violated the principles of American democracy was eventually recognized by the Supreme Court, which required the Pullman Company to divest itself of all property not used for manufacturing purposes. In reaching this decision, the court was in essence enforcing the distinction between different spheres of life. Companies could be owned; towns could not. Property rights extended to commodities, but not to communities.

It certainly seems clear that Pullman had no business owning a town and telling its residents how to live. But how clear would this be if economic imperialism swept aside the barriers between spheres of social life? All Pullman was trying to do was run his town the way he ran his factory. He didn't force anyone to work in his factory, and he didn't force anyone to live in his town. If they did either, however, they were going to play by his rules. What's the difference, after all, between the town and the factory? Pullman owned them both. He had complete control in the factory, by virtue of his ownership. Why not also in the town? We can give no coherent answer to this question unless we keep the economic sphere distinct from the political one and insist that different goods are attainable in the two spheres and different rules govern them.

Indeed, there is a need to keep the economic and political spheres distinct even within the factory. When a man goes to work in a factory, economists argue, he is selling his labor. But it is not just his labor that walks through the factory gate every morning, it is all of him. The laboring man is also the political man, the family man, the consuming man, and the worshiping man.

How much of this man is the employer purchasing with the weekly wage he pays? Is he purchasing the right to tell the man what to wear, or how to comb his hair? Is he purchasing the right to tell the man who to vote for, how to raise his children, or how to treat his wife? Is he purchasing the right to tell the man what religion to worship or what products to buy? Of course not. It seems obvious that he is only purchasing those aspects of the man that are relevant to the successful discharge of his duties.

But it is not so easy to say where relevance stops. A man who doesn't drink off the job will probably do better work on it. A man who keeps himself trim and fit will be sick less often. A man who doesn't cheat on his wife may be more honest and reliable on the job. A man who votes for one candidate may help create a better business climate than a man who votes for the other. A man who buys only the company's products and not those of its competitors will aid the company's profitability. A man who speaks well of the company outside it will encourage others to buy its products. A man who worships a religion that advocates hard work and material achievement will be more productive than a mystic. And a man who raises his children to value material consumption will create more customers for the company in the long run than a man who does not.

It is not very far-fetched to say that all aspects of a man's life, and not just his labor, are relevant to the success of the commercial enterprise he works for. Yet his employer is not allowed to control all aspects of his life. We implicitly acknowledge that when a man goes to work in a particular factory he is choosing a way to make a living not a way to live. We therefore restrict the prerogatives of his employer so that how the man lives can be subject to other sources of control and influence—those of his family, of his government, of his church, and of his own conscience.

Where, then, does the line fall between what an employer does control and what he doesn't? As in the case of all social practices, the answer here is subject to continual disagreement, struggle, reevaluation, and change. There is now a movement afoot, in many different industrial societies, to democratize the workplace. Instead of bringing principles of ownership and commerce into the polity, as Pullman did, workers are attempting to bring principles of democratic politics into the factory. They are seeking a voice in determining when they work, what they make, and how they make it. They are seeking to bring areas of decision authority that have traditionally rested exclusively with ownership onto the shop floor. Just where this movement to democratize the workplace will come to rest is at present an open question. But it depends upon the realization that, even in the factory, not all of the goods that are available are commercial, so that not all of the rules that apply should be the rules that govern the exchange of commodities.

Preserving Spheres of Social Life

Maintaining distinct spheres of life and preventing the emergence of dominant goods is crucial, but it will not be easy, going as it does against the tide of economic imperialism. There are various particular things people can do that would help keep money from becoming society's dominant good. For example, people can attempt to restructure the world of work so that the intrinsic appeal of jobs, high salaries, high status, and power do not all go together. At present, it is largely true that the social value of one's work is directly reflected by one's salary. It needn't be this way. Indeed, society could arrange the highest rates of compensation for the least satisfying jobs, on the presumption that interesting work brings its own rewards. Society would then have doctors, lawyers, teachers, scientists, and engineers who loved medicine, the law, education, science, and technology. Perhaps they would all earn less than automobile workers.

A move like this might seem to be a social disaster. After all, we want the most talented of our citizens entering the professions, and they won't, without the proper incentives. But this line of thinking presupposes that money is the dominant good—the overriding incentive—and it needn't be. If many of the goods that are now available on the market, for money, were instead distributed in some other way, the opportunity to do interesting and challenging work might be incentive enough to attract talented people to the professions. If education, health care, status, and power could not be purchased, people would have less reason than they do now to have monetary incentives dominate their career choices. The result would be people who pursued the professions as practices, for the goods internal to them.

Another move that would help prevent money from becoming a dominant good would be for society to provide for the needs of its citizens *in kind,* not in cash. Food could be provided for the hungry, clothing and housing for the unsheltered, education for the illiterate, and medicine for the sick by means of vouchers, with none of them either transferable or convertible into cash. If this practice were to become widespread, it would become difficult or impossible even to put dollar values on these various social goods. They would be valued in their own terms, and not in terms of a market equivalent.

More generally, though, if people are to keep different spheres of social life distinct, they must be committed to preserving the non-economic character of nonmarket, social institutions. They must resist the pull to think of marriage and family life as a contractual, economic arrangement whose goods can be judged in terms of money equivalents. This is not to say that people must remain loyal to the family as a patriarchal institution, in which the husband and father has ultimate control of all resources and ultimate authority for all decisions. It *is* to say that when people attempt to change the structure of traditional family life, it should be with an eye toward increasing the access of participants to the goods that are appropriate to it and not to economic goods. It is not communal access to stocks, bonds, and bank accounts, but communal access to intimacy, care, and emotional attachment that should be the principal concern.

Similarly, people must preserve the relative autonomy and independence of the school from the market. The goods of education—the provision of critical awareness of self, of others, and of one's social and political institutions—should not be replaced by job training. People must resist the temptation to ask whether college study gives them their money's worth. This is no easy matter, especially as the costs of education continue to skyrocket. But it could be made a good deal easier by, for example, making education—at any level—something that money can't buy.

This discussion of the need to preserve noneconomic institutions that mediate between the individual and the market calls to mind a point discussed in Chapter 6, when we were addressing the limits of economic rationality. Recall that one of the problems raised in that chapter was that whether or not an economic choice is rational will depend on how people keep their accounts, how they decide to keep track of gains and losses. The reason why this poses a problem for economics is that there is no satisfactory theory of how people *ought* to keep their accounts. As with formal accounting, everyday accounting practices are shaped by the goals people have, by the things they are taught, and by the accounting traditions of their culture. Well, insisting that people keep different spheres of life distinct amounts to an argument for a particular way of keeping accounts. People have an education account, a piety account, a family account, a political account, and a morality account—as well as a bank account. And their holdings in each of these accounts cannot—should not—be transferable into any of the others. In this way, social institutions can be preserved, economic imperialism can be stopped, and perhaps economic man and moral man can peacefully coexist.

Stemming the tide of economic imperialism will require action on two fronts. First, people will have to be sure that the social institutions they inhabit are structured so that different spheres of life can retain their distinctiveness. Second, people will have to be sure that the patterns of thinking about themselves—about human nature— that they bring to these institutions encourage them to keep different spheres of life distinct. The objective of this book has been to make a contribution on this second front, by identifying and criticizing an ideology about human nature that threatens to obliterate a host of distinctions on which social life, as we know it, depends.

11

Summing Up

Beware of the man who works hard to learn something, learns it, and finds himself no wiser than before. He is full of murderous resentment of people who are ignorant without having come by their ignorance the hard way.

KURT VONNEGUT

*T*he mission of this book was twofold: first, to review and criticize a picture of human nature that is shared by three seemingly disparate intellectual disciplines, economics, evolutionary biology, and behavior theory; and second, to show that an understanding of these disciplines and of their limitations was not just an academic exercise. Rather the vision that these disciplines share, along with its problems, is reflected in the most fundamental and pervasive institutions and practices that characterize the modern, industrial world. Many of the aspects of modern life that we find most disquieting and troublesome can be traced to the playing out of a particular idea of what it means to be a person that these three disciplines have helped to foster.

Our journey has been slow and laborious. To appreciate the weaknesses of a discipline, one must first apprehend its strengths; to understand what it leaves out of a picture, one must see what it puts in. Although the claims of these disciplines, their limitations, and their influence on our lives are clear, they are not simple. So let us replay the journey, to review where we are and how we got here.

Facts, Morals, and Science

We began by making a distinction between the world of facts and the world of morality—between what *is* and what *ought* to be. This way of carving things up is so familiar that it is hard to imagine an alternative. Yet, although what is does not imply what ought to be in a logical sense, opinions about what ought to be are not independent of what people think is. To develop ideas about right and wrong, we typically depend on some conception, perhaps implicit, of what people are. We have some ideas about human nature—about what people are capable of achieving—and these ideas are what guide us to distinguish right from wrong in particular cases. Moral language is only sensible when applied to full-fledged people, and our understanding of the facts of human nature tells us who the full-fledged people are.

Traditionally, conceptions of human nature have come from many sources—from parents, teachers, priests, everyday experience. But in recent years, all of these sources of knowledge have deferred to the institution that is universally regarded as the last word—the ultimate authority—with regard to facts. That, of course, is the institution of science. Science has told us about the nature of matter, motion, and energy; of digestion, respiration, and reproduction; of ants, birds, and fish. Now we expect it to tell us about the nature of people.

Science as an institution consists of a set of methods used for finding out about the world and a set of criteria used to decide whether these methods really have found things out. We can be said to understand something scientifically when we have discovered the natural laws that govern its activities. Thus, laws of planetary motion, of gravity, and of chemical combination tell us how planets, falling objects, and chemicals must behave. They *must* behave in these ways because it is in their very natures to behave in these ways. Laws of human nature would tell us how people, by their very natures, must behave.

At the same time that science is the most promising source of information about human nature that will help illuminate moral deliberations, it poses a threat to those deliberations. We typically confine moral concerns to those activities where morality makes a difference, activities that involve relations with other people, such as parents, lovers, children, bosses, and customers, where how people act is a matter of discretion—of choice. But suppose there were natural laws governing these activities. Then it might not make much sense to ask how people ought to behave toward others. If, for example, it is human nature to be selfish and dishonest, then what does it matter that we regard selfishness and dishonesty as immoral? We may think it's immoral that planets revolve around the sun also. So what?

There are three scientific disciplines that in the last century have been applying the scientific lens to aspects of human behavior that have traditionally been regarded as moral and discretionary. They are the disciplines of economics, evolutionary biology, and a branch of psychology known as behavior theory. They have in recent years converged on a picture of human nature. It is not a flattering picture. In it, people are painted as greedy, selfish individualists. People are not this way by choice. People are not this way because someone or something made them this way. People are this way by their very natures.

We all know the old aphorism that knowledge is power. Well, like most aphorisms, it isn't entirely true. Knowledge of planetary motion does not give people power over the planets. Knowledge that people are by nature selfish and greedy would not give us power over selfishness and greed. Knowledge gives power only when it identifies aspects of the world over which people can exercise some control. If the economists, biologists, and psychologists are right, selfishness and greed are not characteristics of people that can be controlled or altered. We simply have to learn to live with them, to make the best of an imperfect situation. Are the economists, biologists, and psychologists right?

The bulk of the book has been devoted to answering this question. And the answer I have suggested is no. It is not that people aren't selfish and greedy, for perhaps they are. Rather, it is that they aren't selfish and greedy *by nature*. The social conditions in which people live can make them selfish and greedy—can make selfishness and greed the only games in town. But social conditions can be altered and, with them, selfishness and greed. The disciplines in question have mistakenly treated the particular social conditions in which we live as representative of the universal human condition. As a result, they have mistaken local cultural and historical truths about people for natural laws. As a further result they have helped contribute to the perpetuation of these conditions by appealing to their natural inevitability.

Arriving at this answer required a review of the principles of human nature developed by economics, biology, and behavior theory, and of the evidence from which those principles were derived, as well as a discussion of the limitations of these disciplines, of the reasons we have for regarding their formulations with suspicion. Let us first summarize what these three sciences have to say about human nature.

Economics, Biology, and Behavior Theory

The modern science of economics began with Adam Smith in the eighteenth century. It was, and is, concerned with answering questions about the allocation of scarce resources. Scarcity posed problems long before Adam Smith. Before Smith, societies managed scarcity by appealing to their traditions, traditions that prescribed certain kinds of human conduct and proscribed others. Smith offered to replace tradition with science. He assumed that societies were just collections of individuals, each of whom was out to satisfy his economic interests. He argued that the fundamental economic act was exchange and that the fundamental economic value was exchange value, and he suggested that if we simply left people alone to pursue their interests, by engaging in exchanges with one another in a free and competitive marketplace, the activities of each would be guided, as if by an invisible hand, to secure the well-being of all. No scheme of resource allocation could be developed that was more efficient and effective than the free and selfish pursuit of individual economic interest in the market. If the market were truly free, with supply and demand interacting to determine the price of land, of labor, and of goods, rational economic men would succeed in getting what they wanted. And if individuals got what they wanted, so would society as a whole, since it was nothing more than a collection of these individuals.

Since Smith's time, the notion of "rational economic man" has been refined, elaborated, and formalized. To the modern economist, economic rationality includes these features:

1. The things that people value are individual, idiosyncratic, and incommensurable.
2. People always value and want something.
3. People can express preferences among all commodities.
4. They prefer more of something to less.
5. They prefer low prices to high ones.
6. They have relatively stable preferences.
7. They have transitive preferences.
8. Their preferences obey the law of diminishing marginal utility.
9. They choose with complete information.
10. They always choose that set of goods available that maximizes their preferences, subject to the constraints of their resources.

Smith did not suppose that this rational, economic piece of people was all there was to human nature. In addition to rational economic man, there was moral man, who took the time out from the pursuit of his interests to show concern and sympathy for his fellows. Moral man was the relevant actor on the stage of the family, the church, and the state. Indeed, this moral side of man even played an important role in keeping the market together as an institution. But within the market it was not concern or sympathy, but economic self-interest, that governed the behavior of the players.

Enter biology, or more specifically, the part of evolutionary biology known as sociobiology. What sociobiology has done is extend the conception of economic rationality to domains of life that have been excluded by most economists—domains including social relations within a group, relations between parents and offspring, and relations between mates. The origins of sociobiology are in Charles Darwin's theory of evolution by natural selection. Having read Malthus, an economist contemporary of Adam Smith's who saw doom rather than prosperity in the mirror of the market, Darwin reasoned that organisms were in steady competition for nature's scarce resources. The ones we observed were well suited to their environments not because anyone had designed them that way. Rather, they were the ones among many essentially randomly constructed contestants who had survived and reproduced. All the creatures we saw were adapted to their environments because the ones that hadn't adapted had died off. Survival of the fittest. Evolution by natural selection. Another invisible hand seeing to it that only organisms that were able to cope survived.

Modern evolutionary biology has elaborated this Darwinian insight to account for the social behavior of animals. The animal's biological "interest" is in reproduction. After all, the only way to survive over periods of evolutionary time is to put copies of oneself into the next generation. And not even of oneself, but of one's *genes,* those information-rich packets of protein that are the vehicle for transmission of traits from parent to offspring—from one generation to the next.

Modern evolutionary biology is the biology of selfish genes, entities whose only concern is their own proliferation. Such complex and seemingly unselfish activities as mating, caring for the young, and warning the group about potential predation can all be understood as serving the interests of selfish genes. Acts of self-sacrifice are explained by appealing to the ideas of inclusive fitness and kin selection; an individual's genes are not just in it, but in its children, its brothers, its aunts, and its cousins. So it is in the interests of selfish genes to promote the survival of all relatives. But all is selfishness underneath. "Scratch an altruist and watch a hypocrite bleed."

Concepts from sociobiology have been used to explain patterns of human social behavior in the same way they explain the social behavior of other animals. The promiscuity of men and the fidelity of women, the extreme devotion of women to child care, the extraordinary selectivity of women relative to men in choice of mate—all of these characteristics of human social behavior and more can be traced to the action of selfish genes. And so, in the hands of sociobiology, the conception of people as in slavish pursuit of self-interest extends beyond the bounds of the market to virtually all aspects of social life. Furthermore, human selfishness is clearly a reflection of natural law, since it is of a piece with the selfishness of ants, birds, fish, and all other living organisms.

Sociobiology emphasizes patterns of behavior that characterize all members of a species and are the relatively inflexible product of genetic determination. But a hallmark of human behavior seems to be its diversity and flexibility, its susceptibility to modification by experience. Even if the wealth of different patterns of social behavior that people have displayed in different places and at different times merely masks the universal pursuit of reproductive success by selfish genes, the diversity itself must be accounted for. And sociobiology can't provide the account. Enter the last of our scientific disciplines, behavior theory, to provide the final piece in the puzzle of human nature.

Behavior theory offers a mechanism to explain the development of adaptive patterns of behavior within the life of an individual that exactly parallels the mechanism of natural selection in the evolutionary development of a species. The mechanism is the principle of reinforcement. Organisms engage in varied, essentially random patterns of behavior. Some of that behavior has favorable consequences; it results in states of the environment that organisms want, that are reinforcing. The behavior that results in reinforcing consequences, and only that behavior, continues to occur; other, less successful behavior drops out. Because of this natural selection by reinforcement of behavior that works, the organisms that we see have learned to do just the right things to produce the outcomes they want or need. So just as the maximization of reproductive success drives the evolution of species, the maximization of reinforcement drives the development of individuals.

To test the power of the principle of reinforcement, behavior theorists study rats and pigeons in laboratory environments in which arbitrary bits of behavior, like lever presses or pecks, are the means to reinforcers like food or water. And the principle is powerful indeed:

behavior can be manipulated and controlled in exquisite detail by manipulating the relation between that behavior and reinforcement. Furthermore, the principle of reinforcement proves powerful with people as well when it is used to control behavior in applied settings like mental hospitals, schools, and workplaces.

This picture of organisms as pursuers of reinforcement maximization fits nicely with the model of rational economic man. Indeed, the research of behavior theorists in recent years has suggested that the phrase "rational economic man" be revised to "rational economic organism." When pigeons and rats are required to choose how to allocate their scarce resources (behavior) over a range of available commodities (reinforcers), they choose in just the way that economists say people do. Economic variables like relative price, demand elasticity, substitutability, and temporal discounting seem to affect the choices of pigeons just as they affect the choices of people. Behavior theory is fast becoming a kind of experimental economics.

When we add these three disciplines together, they converge on a picture of self-interested, acquisitive human nature that is truly formidable. Moreover, each discipline provides a weapon to be used by the others against the sorts of criticisms most commonly directed at them. The economist says that people are, by their very natures, self-interested and greedy. The critic says no, that greedy economic men are made, not born. The pursuit of self-interest is the only option in a free-market society. The economist can now turn, in defense, to sociobiology. Ants, birds, and fish don't live in artificially created, free-market societies, and they pursue self-interest also. The sociobiologist can in turn be criticized for his emphasis on genetic determination of behavior, an emphasis which flies in the face of the self-evident facts of human flexibility and diversity. He can now turn to the behavior theorist for evidence that flexibility and diversity are themselves governed by principles of self-interest (reinforcement) maximization. Finally, the behavior theorist can be criticized for finding principles that are true of pigeons in simple laboratory environments but are not true of people in real life. He can turn to the economist, whose principles are about people in real life, and point to the convergence of behavior theory and economics as a validation of behavior theory. The fit between these disciplines is neat, almost seamless. The temptation is strong to stop treating greedy self-interest as a moral matter and accept it as a matter of fact.

The Limits of Science

Despite this impressive convergence, as we examined each of the three disciplines carefully, we uncovered problems or limitations with each that could not be addressed by contributions from the others. While the view that people are ruled by greedy self-interest must be taken seriously, it is false. Or at least it is false if it is taken to be an eternal law of human nature. The problems fell into two general classes. First, each discipline is importantly incomplete or inaccurate even within its own relatively narrowly defined domain. And second, even if we accept what the disciplines have to say within their own domains, there is no reason to accept their principles as a general account of what people are. Let us review these two classes of problems separately.

Economists claim that people act in the market as rational economic agents, with economic rationality identified by a list of precise and detailed characteristics. There is good reason to believe, however, that people violate each and every one of the features of economic rationality in at least some important circumstances. People cannot always express preferences among commodities, they don't always prefer lower prices to higher, they rarely choose with perfect information, their preferences are not always transitive, nor are they stable, and they rarely choose so as to maximize preference. Further, even to begin to judge whether people are behaving rationally, we have to know how they keep their personal accounts of costs and benefits, and economics has nothing to offer about this important matter. Economists can model the behavior of rational economic agents under idealized, abstract conditions, but real life is neither idealized nor abstract, and the conditions of real life make a big difference.

Sociobiologists argue that organisms act to maximize inclusive reproductive fitness, engaging in activities that will propel the greatest possible number of their genes into the next generation. When we evaluate this claim empirically, the evidence is clear that fitness maximization never occurs. Organisms possess traits that are neutral with regard to selection but that nevertheless constrain the evolution of other traits. There are time lags between evolutionary change and environmental change so that when a trait first appears it may be wonderfully adaptive, but by the time it has come to dominate a species it may no longer be so adaptive because the environmental conditions that made it so adaptive have altered. The adaptiveness of a trait depends upon the genetic and behavioral context in which it

occurs. There are historical constraints on maximization; each evolutionary change along the way from point A to point Z must itself be
adaptive. And there are constraints on the range of variation from
which natural selection has the opportunity to select. The move
sociobiologists make is to claim that evolution maximizes fitness subject to various background constraints on perfection like these. But
they never indicate what is to count as a fitness-maximizing trait and
what is to count as a background constraint. The result is that a
fitness-maximizing story can be told about virtually anything.

Finally, there is behavior theory. Behavior theory tells us that an
organism's behavior is shaped by the principle of reinforcement, that
the selection of behavior is reinforcement maximizing. This claim
turns out to be empirically true of rats and pigeons only when they
are studied in environments that prevent behavior that is not reinforcement maximizing from occurring. When environments are
enriched sufficiently to allow influences other than reinforcement to
be felt, organisms frequently "misbehave"; they engage in activities
that actually *cost* them reinforcement. Thus, it is fair to say that
while reinforcement *can* control the behavior of organisms under certain highly restricted conditions, it need not ordinarily or character-
istically do so.

Now what if we treat these various limitations as slight imperfections in immature sciences that will eventually be cleared up? What
if we grant that people are rational economic agents, that natural
selection does work to maximize the inclusive fitness of animals, and
that pigeons and rats are in general governed by the principle of reinforcement? Do these concessions imply that we accept greedy self-
interest as the principle that characterizes human nature? No, they
do not.

For economics has a second problem. Formulations about economic
rationality presuppose the existence of markets for labor and commodities and assume that the fundamental economic act is exchange,
rather than, say, production. But the work of economic anthropologists makes it clear that the free and autonomous market is a recent
human invention. At other times in history (and even at present in
some cultures), there weren't markets as we know them today. Labor
was not sold to the highest bidder, and acts of exchange were not
motivated by the desire for gain. Although there was economic activity, to be sure, it was activity that was embedded in other social institutions and not given free reign in an autonomous market. The market

system in its present form is in part the product of Adam Smith and his contemporaries.

What economics has done is take patterns of behavior that are produced by one particular kind of social arrangement—the market—and elevate them to the status of eternal and universal principles of human nature. That economic rationality depends on the existence of markets and that markets may or may not be a part of human society tells us something about the status of economic laws as laws of human nature. The laws of economics, unlike, say, the law of gravity, can be broken, by creating social institutions that do not have the properties of the market. So the slavish pursuit of self-interest is a matter of human discretion. We can ask whether people *should* pursue self-interest to the degree that they do, and if we decide they should not, we can do something about it by altering or eliminating the market as an institution. In short, whether there should be markets, how they should run, and what should be available in them are all questions whose answers must come from the world of morality, not the world of facts.

And sociobiology also has a second problem. Even if we concede that it is a powerful framework for understanding the behavior of animals, what does it tell us about people? The answer may be precious little. In accounting for animal behavior, the sociobiologist assumes, for good reason, that there is a tight link between motive and action. Both the ends an organism seeks and the means it uses to achieve those ends are tightly connected to each other and to the genes that largely govern them. But in the case of people, there is no such inflexible link. Both motives and actions display enormous diversity across people and across cultures. What is significant about actions is typically not their physical character but their meaning, and the meaning of actions is largely culture-determined and culture-specific.

Even kinship, a concept that is absolutely essential to the sociobiological story, means different things in different cultures; only rarely does it mean what sociobiologists assume it must mean. What is striking about people is not how flexible and intelligent they are, but how incomplete, unfinished, and culture-dependent that flexibility makes them. The rules of conduct that are operative in a culture, and the meanings that the culture gives to actions, may be in agreement with the principles of sociobiology, and when they are, the principles of human nature offered by the sociobiologists will gain a measure of confirmation. But cultural rules and biological laws need not be in agreement. Whether they are depends upon how people decide to organize their culture. It is thus appropriate to ask whether people should be inclusive fitness maximizers, for it clearly isn't true that they must be. So as in the case of economics, whether people conform to the sociobiological vision of human nature is a question of moral, not of natural imperatives.

We come finally to behavior theory, and of it we ask, what reason is there to believe that even if it is true of rats and pigeons, it is also true of people? The behavior theorist can reply by pointing to successful applications of the principle of reinforcement in human settings. But this reply won't do, for just as we now know that the behavior of pigeons can be made to conform to the principle of reinforcement by eliminating other influences from the situation, so can the behavior of people. This tells us little about the role of reinforcement in everyday settings that have not been carefully created by the behavior theorist.

The behavior theorist can point to the modern workplace as an exemplification of the principle of reinforcement in a situation he had no hand in creating. But as we saw, like the free market, the modern workplace is a recent historical invention. It has been created over the last two hundred years with an explicit eye toward eliminating influences on work aside from reinforcement from exerting themselves. Once, social custom played a large role in determining what work a person did, how he did it, who he did it for, and what he was paid for it. The modern workplace has detached work from custom by breaking it down into meaningless, repetitive units and separating it from the rest of a person's life. There is nothing left to influence work aside from the reinforcement it brings. Small wonder that reinforcement is so effective.

It is not hard to produce this transformation in the nature and control of work, as we saw from the laboratory demonstration that turned inquisitive, problem-solving college students into stereotyped assembly-line workers. That the modern workplace is a recent human invention tells us the same thing about behavior theory that we have already learned about sociobiology and economics. Work could be organized in this way, in which case it would conform to behavior theory's description of human nature. But it could be organized some other way. Which way it is organized is a matter of human discretion. The question is how work *should* be organized. It is a question of moral evaluation, not natural imperative.

Economic Imperialism

Why is it so important to understand these three disciplines and their limitations? Why should we struggle to know that what they take to be matters of natural necessity are really matters of human social discretion? The society we inhabit has taken the message that these disciplines offer as its implicit ideology. Increasingly, aspects of life that were never before regarded as matters of economic self-interest are coming to be modeled on the market. And this phenomenon of "economic imperialism" poses a dire threat to many of our social institutions, including the institution of the market itself.

First, there are many situations in which the rational pursuit of individual interests ends up making everyone worse off. Phenomena like the problem of the commons and the prisoner's dilemma are model cases in which individual goods produce collective bads. What makes these problems so difficult to solve is that it is not economically rational for any individual to deflect from the pursuit of his interests and pursue instead the common good.

One can solve these problems by coercing people to act for the common good, or one can solve them by appealing to the moral rather than the economic side of people. But both of these solutions are problematic themselves. Coercion entails a loss of freedom. On top of that, to coerce people requires that we have the power to enforce the social sanctions we impose. But enforcement will be impossible if everyone is prepared to ignore social sanctions if he thinks he can get away with it. In short, coercion can only work if most people will obey the law even if it is not in their individual economic interest to do so. We can't count on rational economic agents to obey the law. The same problem applies to attempts to appeal to the moral side of people. As the market extends to aspects of life that were previously noneconomic—like marriage, childrearing, and education—where do we expect people's moral resources to come from? Markets undermine morals, and as they do so, appeals to anything other than self-interest can be expected to fall on increasingly deaf ears.

Aside from making it harder and harder to achieve collective goods, economic imperialism threatens to undermine various significant social practices by turning the goals that direct them from noneconomic into economic ones. The activities of doctors, lawyers, teachers, scientists, professional athletes, and even husbands and wives look very different when they are guided by the pursuit of gain than when they are guided by the pursuit of healthy people, just resolutions of conflict, enlightened children, truths about nature, excellence in physical accomplishment, and intimate emotional bonds. And as these various practices erode, it will become increasingly difficult to find anything other than economic interest as a reason for action.

One ironic consequence of this single-minded pursuit of economic interest will be that the market itself will cease to function effectively. For in order for the market to work, people must make moral commitments to agree on what can be bought or sold, to tell the truth, and to honor their contracts. In the absence of these commitments, the market will grind to a halt. And as the market grinds to a halt, so will our system of political democracy. Unless people are willing to submerge their individual interests at least some of the time for the common good, we will not be able to afford democracy any longer. Significant aspects of political life will have to be taken out of our hands and left to be run by some powerful institution that is invulnerable to individual political preferences.

What can be done to forestall these developments? The economist, the sociobiologist, and the behavior theorist might say nothing. They might say that what we are seeing is just the playing out of natural human impulses. But we now know that they are wrong. These developments can be reversed. We can choose to protect significant domains of social life from economic imperialism. We can insist that institutions like the family, the church, the school, and the state be guided by goals that are not economic. We can insist that the market be restricted to a very narrow domain. We still have the power to insist on these things. The question is whether we still have the sense to.

Epilogue

*So far as I am aware, we are the only society that thinks of
itself as having risen from savagery, identified with a ruthless
nature. Everyone else believes they are descended from gods.
. . . Judging from social behavior, this contrast may well be
a fair statement of the differences between ourselves and the
rest of the world. We make both a folklore and a science of
our brutish origins, sometimes with precious little to distinguish between them.*

MARSHALL SAHLINS

*A**nimals* are born into wilderness. They walk
on paths that are accidents of nature. There is no plan to the paths—
no organization. Some paths lead nowhere. Others turn back on
themselves. Some are destroyed by floods, others opened by fire. Different animals stumble along different paths as far as the paths will
take them. They don't know where they are going, or even that they
are going anywhere.

People are born into culture. They walk on paths that were constructed by their ancestors. The paths are protected by walls that
keep the wilderness away. In some cultures, there are only a few
well-marked paths to travel; all the rest is jungle. And the paths do
not intersect. In other cultures, there are many paths. some are
superhighways, built to accommodate many, moving at high speed.
Others are just a single lane, built for only a few. Still others are just
dirt roads, traveled by an occasional stray foot. These paths intersect

in many places, allowing travelers to get off one and onto another. As culture develops, the paths are changed. Some stop being used and are allowed to fall into disrepair, slowly reclaimed by the wilderness. Others become popular and are lengthened and expanded to make room for all travelers. Culture's paths are not accidental. They are meant to constrain people to move in some directions and not others; to make some destinations easy to reach and others impossible. These paths are meant to help travelers find their way.

In this book, we have toured one particular path, the path of rational economic man. Its construction began three hundred years ago. At the time it began, it led in a direction that it occurred to hardly anyone to travel. People had to be cajoled to explore it. "I can see where this path shall lead," said Adam Smith. "There is gold at the end for everyone." Still, people were reluctant. So slowly, other paths were closed and people were forced onto this one. The industrial revolution widened the path of economic man and paved it. Some still disdained the path. "People who walk down this path will come to no good," they said. "It is strewn with corruption; it will lead us back into the wilderness. It is not the path of God."

"No," said the sociobiologist, "this is the path of nature. This is the path we have always been meant to be on." And the behavior theorist agreed. "All paths lead to the same place," he said. "Our old ones are littered with superstitions, with false beliefs that serve as detours and roadblocks. This new path takes us where we were going anyway, but it takes us quickly, efficiently, and directly."

So more people entered the path. It was widened still further. And those who stubbornly refused to leave their old, familiar paths were fooled. While they were not being vigilant, their paths were rerouted, by economic imperialism, so that they ultimately joined the main one. Though these people thought they were pursuing destinations that were different from the destination of economic man, they were wrong. In the end, there was only one destination left to pursue. The old paths only got there more slowly. Seeing this, the last holdouts joined the main path also, resigned to the fact that there was nowhere else to travel.

But with everyone now on the path, moving single-mindedly and at high speed, a problem has arisen. There is no one left to maintain the path. No one is willing to slow down and repair potholes. No one wants to spend time widening the path so that it will accommodate all its traffic more smoothly. No one wants to carpool, or to pick up broken down travelers stranded by the side of the road. Cracks are appearing in the path—big ones, with weeds that are harbingers of encroaching wilderness growing through them. More and more, people are being forced to stop; they are running out of gas. Their heavy vehicles are destroying the road and at the same time being destroyed by it.

Perhaps if the deterioration continues, people will try to get off this path before the wilderness takes it over. Perhaps they will try to reopen some of the old ones before they are so overgrown that they can't even be found. If they don't do it soon, they will have to start from scratch, hacking out new paths with no assurances as to where they will lead.

Rational economic man as a reflection of human nature is a fiction. It is a modern invention, a new path. But it is a powerful fiction. and it becomes less and less a fiction as more and more of our institutions get pervaded by its assumptions and other paths are closed. Because of its self-fulfilling character, we can't expect this fiction to die of natural causes. Clifford Geertz has said of this fiction, "It is the moral equivalent of fast food, not so much artlessly neutral as skillfully impoverished." But to see it as impoverished, we must be able to hold onto the alternatives. We must keep the other paths open. And this will not be easy.

Notes

S*ince* this is not a scholarly monograph, detailed documentation of sources seems inappropriate. Instead, for each chapter, I have provided a few references that should be both interesting and accessible to the general reader. These are works that have influenced my own thinking, and that can give the reader more information than I have had space to provide about topics of special interest or importance. In addition to these general references, I have provided full bibliographic information for each work mentioned in the book, and for each quotation. This information is keyed to the page on which the reference or quotation appears.

Chapter 1

My discussion of modern American individualism and the malaise it seems to be engendering is taken from R. N. Bellah, R. Madsen, W. M. Sullivan, A. Swidler, & S. M. Tipton, (1985). *Habits of the Heart* (Berkeley: University of California Press, 1985). Their discussion, in turn, depends heavily on two works. One is classic: A. de Tocqueville, *Democracy in America* (1835, 1840: New York: Doubleday, 1969). The other is modern: A. MacIntyre, *After Virtue* (South Bend, In.: University of Notre Dame Press 1981).

page 15. The proverb is taken from R. Ruark, *Something of Value* (New York: Random House, 1953).

page 17. Bellah, *Habits of the Heart,* pp. 298–299.

page 18. R. Frost, "The Death of the Hired Man" (1914).

pages 19. Bellah, *Habits of the Heart.* The phrase "community of memory" is theirs, as is the notion of America's "first" and "second" languages.

page 19. De Tocqueville, *Democracy in America.*

page 19. De Tocqueville, *Democracy in America,* p. 292.

page 20. Bellah, *Habits of the Heart,* pp. 219, 324.

page 21. Aristotle, *Nicomachean Ethics,* Books 7 and 9.

page 22. J. Winthrop, "A Model of Christian Charity" (1630).

Chapter 2

page 23. J. M. Keynes, *The General Theory of Employment, Interest, and Money.* (New York: Harcourt, 1936), p. 532; M. Kundera, The novel and Europe. *New York Review of Books* 31(14) (1984): p. 15.

page 25. D. Hume, *A Treatise of Human Nature* (1739).

page 28. M. Kundera, *The Unbearable Lightness of Being* (New York: Harper & Row, 1984).

page 41. T. Hobbes, *Leviathan* (1651).

page 44. C. Darwin, *The Origin of Species* (1859).

page 46. R. Hofstadter, *Social Darwinism in American Thought* (New York: Braziller, 1959), p. 57.

page 47. A. Smith, *The Wealth of Nations* (1776).

page 47. Hofstadter, *Social Darwinism in American Thought,* p. 144.

Chapter 3

Our discussion in this chapter was more concerned with the foundations than with the details of economic theory, and most books written by economists are more concerned with the details than the foundations. Thus, the books recommended here were not all written by economists. For a clear, elementary discussion of the foundations of economics, see C. Dyke, *Philosophy of Economics* (Englewood Cliffs, N.J.: Prentice-Hall, 1981). A somewhat more sophisticated discussion can be found in M. Hollis and E. Nell, *Rational Economic Man* (Cambridge: Cambridge University Press, 1975). For still more sophistication, see T. Koopmans, *Three Essays on the State of Economic Science* (New York: McGraw-Hill, 1957). A very readable account of the history of economic thought is R. L. Heilbroner, *The Worldly Philosophers,* 5th ed. (New York: Simon and Schuster, 1980). Another historical story, on how the notion of "interest" was born, is told by A. O. Hirschman, *The Passions and the Interests* (Princeton: Princeton University Press, 1977). Finally, for a discussion of the details, a clear, elementary text is E. Mansfield, *Principles of Microeconomics,* 5th ed. (New York: Norton, 1986).

page 54. G. Becker, *The Economic Approach to Human Behavior* (Chicago: University of Chicago Press, 1976), p. 8; F. Edgeworth, *Mathematical Psychics* (London: Kegan-Paul, 1881), p. 16.

page 57. A. Smith, *The Wealth of Nations* (New York: Random House, 1937), p. 119.

page 63. Smith, *The Wealth of Nations,* pp. 4–5.

page 64. A. Smith, *The Theory of Moral Sentiments* (1753; (Oxford: Clarendon Press, 1976), pp. 124–125.

page 65. Smith, *The Wealth of Nations,* pp. 734–735.

page 65. A. Marshall, *Principles of Economics,* 8th ed. (London: Macmillan, 1920), pp. 90–91.

page 73. S. Jevons, *Theory of Political Economy* (New York: A. M. Kelly, 1871), p. 9.

page 75. Smith, *The Wealth of Nations,* p. 14.

page 77. G. Becker, E. Landes, and R. Michael, An economic analysis of marital stability, *Journal of Political Economy* 85 (1977): 1143.

page 81. J. S. Mill, *Principles of Political Economy,* vol. 2 (New York: Columbia University Press, 1899), p. 276.

page 84. R. Nozick, *Anarchy, State, and Utopia* (New York: Basic Books, 1974).

page 84. J. Rawls, *A Theory of Justice* (Cambridge: Harvard University Press, 1971).

Chapter 4

Three nontechnical accounts of the new evolutionary biology worth reading are R. Dawkins, *The Selfish Gene* (Oxford: Oxford University Press, 1976); D. Barasch, *The Whisperings Within* (New York: Harper & Row, 1979); and E. O. Wilson, *On Human Nature* (Cambridge: Harvard University Press, 1978). More technical discussions are Wilson's *Sociobiology* (Cambridge: Harvard University Press, 1975), which started it all, and Barasch's *Sociobiology and Behavior,* 2nd ed. (New York: Elsevier, 1982). A good account of the origins of Darwinian theory is M. Ruse, *The Darwinian Revolution* (Chicago: University of Chicago Press, 1979).

page 85. M. Ghiselin, *The Economy of Nature and the Evolution of Sex* (Berkeley: University of California Press, 1976), p. 247; S. Bellow, cited in D. Symons, *The Evolution of Human Sexuality* (Oxford: Oxford University Press, 1979).

page 89. C. Darwin, *The Origin of Species* (1859; Cambridge: Harvard University Press, 1964). pp. 63–64.

page 90. C. Darwin, *The Descent of Man* (London: Murray, 1871).

page 91. P. Medawar, cited in G. C. Williams, *Adaptation and Natural Selection* (Princeton: Princeton University Press, 1966), p. 158.

page 115. Dawkins, *The Selfish Gene,* p. 3; Barasch, *The Whisperings Within,* p. 90.

Chapter 5

For a sophisticated but readable discussion of the fundamental assumptions, commitments, and principles of behavior theory, there is no substitute for reading B. F. Skinner, the intellectual leader of the discipline. In *Science and Human Behavior* (New York: Macmillan, 1953), he presents the basic principles in nontechnical language and extrapolates them to countless situations in daily life. In *Beyond Freedom and Dignity* (New York: Knopf, 1971), he presents a challenging alternative to our everyday conception of human nature that follows from the fundamental principles of behavior theory. In *About Behaviorism* (New York: Knopf, 1974), he discusses the methodological commitments that behavior theory makes in striving to attain a science of human nature. For discussions of basic principles and applications of behavior theory that are more up to date and detailed than Skinner's, see B. Schwartz and H. Lacey, *Behaviorism, Science, and Human Nature* (New York: Norton 1982); or, for a more sophisticated treatment, B. Schwartz, *Psychology of Learning and Behavior*, 2nd ed. (New York: Norton, 1984).

page 117. B. F. Skinner, Selection by consequences. *Science* 213 (1981): p. 501; J. E. R. Staddon, *Limits to Action* (New York: Academic Press, 1981), p. xviii.

Chapter 6

Among the books mentioned in the notes to Chapter 3, Koopmans' *Three Essays on the State of Economic Science,* Dyke's *Philosophy of Economics,* and Hollis and Nell's *Rational Economic Man* provide critical discussions of the foundations of economic theory. In addition, see T. Scitovsky, *The Joyless Economy* (Oxford: Oxford University Press, 1976); A. O. Hirschman, *Shifting Involvements* (Princeton: Princeton University Press, 1982); and A. K. Sen, *Collective Choice and Social Welfare* (San Francisco: Holden-Day, 1970). Finally, for a detailed argument for the historical specificity of the free-market system, see K. Polanyi, *The Great Transformation* (New York: Rinehart, 1944).

page 152. K. Arrow, *The Limits of Organization* (New York: Norton, 1974), p. 27; Polanyi, *The Great Transformation,* p. 46.

page 157. T. Veblen, *The Theory of the Leisure Class* (1899; New York: Modern Library, 1934).

page 161. H. Simon, Rational choice and the structure of the environment. *Psychological Review* 63 (1956): 129–137.

page 162. Plato, *Phaedo* (New York: Liveright, 1927), p. 52.

page 162. R. L. Solomon, The opponent process theory of acquired motivation. *American Psychologist* 35 (1980): 691–712.

page 164. Scitovsky, *The Joyless Economy.*

page 166. A. Smith, *The Theory of Moral Sentiments* (Oxford: Clarendon Press, 1753), p. 302.

page 167. A. Tversky and D. Kahneman, The framing of decisions and the psychology of choice, *Science* 211 (1981): 453–458. Also, D. Kahneman and A. Tversky, Choices, values, and frames, *American Psychologist* 39 (1984): 341–350.

page 174. A. Camus, cited in J. O'Toole *Work in America* (Cambridge: MIT Press, 1973), p. 186.

page 174. A. Smith, *The Wealth of Nations* (New York: Random House, 1937), pp. 734–735.

page 176. Polanyi, *The Great Transformation*, pp. 43–44.

page 176. M. Sahlins, *Stone Age Economics* (Chicago: University of Chicago Press, 1967).

page 176. Aristotle, *Politics*.

page 179. Polanyi, *The Great Transformation*, p. 30.

page 180. Dyke, *Philosophy of Economics*.

Chapter 7

Discussions of the problems and limits of theorizing in evolutionary biology can be found in E. Sober, *The Nature of Selection* (Cambridge: MIT Press, 1984); and in J. Elster, *Ulysses and the Sirens* (Cambridge: Cambridge University Press, 1979). Critical discussions of biological accounts of human social behavior that emphasize the importance of culture to human life are C. Geertz, *The Interpretation of Cultures* (New York: Basic Books, 1973); M. Sahlins, *The Use and Abuse of Biology* (Ann Arbor: University of Michigan Press, 1976); and K. Bock, *Human Nature and History* (New York: Columbia University Press, 1980). A detailed and decisive critique of the entire sociobiological project has recently been provided by P. Kitcher, *Vaulting Ambition* (Cambridge: MIT Press, 1985).

page 182. C. Geertz, in J. R. Platt, ed., *New Views on the Nature of Man* (Chicago: University of Chicago Press, 1965), p. 63; M. Sahlins, *Culture and Practical Reason* (Chicago: University of Chicago Press, 1976), p. 208; C. Darwin, *The Voyage of the Beagle* (1839), cited in S. J. Gould, *The Mismeasure of Man* (New York: Norton, 1981).

page 185. R. Dawkins, *The Extended Phenotype* (San Francisco: Freeman, 1982).

page 192. Symons, *The Evolution of Human Sexuality*, p. 202.

page 198. Elster, *Ulysses and the Sirens*.

page 205. R. Axelrod, *The Evolution of Cooperation* (New York: Basic Books, 1984).

page 209. R. D. Alexander, The search for a general theory of behavior, *Behavioral Science* 20 (1975): 96.

page 210. Sahlins, *The Use and Abuse of Biology*.

page 211. B. Malinowski, *The Sexual Life of Savages in North-Western Melanesia* (New York: Eugenics Press, 1929), p. 193.

page 212. Sahlins, *The Use and Abuse of Biology,* p. 8.
page 213. C. Geertz, *The Interpretation of Cultures,* p. 64.
page 213. C. Geertz, *The Interpretation of Cultures,* p. 46.

Chapter 8

Discussions of the changes in the nature of work ushered in by the industrial revolution can be found in K. Polanyi, *The Great Transformation* (New
York: Rinehart, 1944); in H. Braverman, *Labor and Monopoly Capital* (New
York: Monthly Review Press, 1964); and in E. J. Hobsbawm, *Labouring Men*
(London: Weidenfeld and Nicholson, 1964). There is no single nontechnical
book about the biological influences on the behavior of animals in operant
conditioning experiments, but more detail can be found in Schwartz, *Psychology of Learning and Behavior,* 2nd ed. (New York: Norton, 1984), ch. 12.

page 216. M. Sahlins, *Culture and Practical Reason* (Chicago: University of
Chicago Press, 1976), p. 216.
page 221. K. Breland and M. Breland, The misbehavior of organisms, *American Psychologist* 16 (1961): p. 682.
page 222. Breland and Breland, The misbehavior of organisms, p. 683.
page 223. B. F. Skinner, Herrnstein and the Evolution of Behaviorism,
American Psychologist 32 (1977): 1007.
page 229. Hobsbawm, *Labouring Men,* pp. 348–349.
page 231. F. W. Taylor, *Principles of Scientific Management* (1911; New York:
Norton, 1967), pp. 24–25.
page 232. L. M. Gilbreth, *The Psychology of Management* (New York: Sturgis
and Walton, 1914), p. 292.
page 232. C. B. Ferster and B. F. Skinner, *Schedules of Reinforcement* (New
York: Appleton-Century-Crofts, 1957).
page 232. Braverman, *Labor and Monopoly Capital,* p. 124.
page 232. P. F. Drucker, *The Practice of Management* (New York: McGraw-
Hill, 1954), p. 280.
page 234. S. Marglin, What do bosses do? In *The Division of Labour* ed. A.
Gorz (London: Harvester Press, 1976), pp. 13–54.
page 235. A. Smith, *The Wealth of Nations* (New York: Random House,
1977), p. 7.
page 238. A. Ure, *The Philosophy of Manufactures* (London: Charles Knight,
1835), pp. 15–16.
page 238. Marglin, What do bosses do? pp. 38–39.
page 240. M. R. Lepper and D. Greene, eds. *The Hidden Costs of Reward*
(Hillsdale, N.J.: Erlbaum, 1978).
page 241. B. Schwartz, Reinforcement-induced behavioral stereotypy: How

not to teach people to discover rules, *Journal of Experimental Psychology: General 111 (1982)*: 23–59.

page 246. Smith, *The Wealth of Nations*, pp. 734–735.

Chapter 9

The logic by which the pursuit of individual interests fails to secure the collective interest, and may even interfere with it, is discussed by A. K. Sen, in *Collective Choice and Social Welfare* (San Francisco: Holder-Day, 1970), and by M. Olson, *The Logic of Collective Action* (Cambridge: Harvard University Press, 1965). The problems of positional goods and economic imperialism are developed in detail by F. Hirsch, *Social Limits to Growth* (Cambridge: Harvard University Press, 1976). Practices and virtues are discussed by A. MacIntyre in *After Virtue* (South Bend: University of Notre Dame Press, 1981).

page 247. C. L. Schultze, *The Public Use of the Private Interest* (Washington D.C.: Brookings Institution, 1977), p. 18.

page 252. Hirsch, *Social Limits to Growth*.

page 252. L. Thurow, *The Zero-Sum Society* (New York: Basic Books, 1980).

page 254. T. Schelling, *Micromotives and Macrobehavior* (New York: Norton, 1978).

page 255. G. Hardin, The tragedy of the commons. *Science* 162 (1968): 1243–1248.

page 257. Schelling, *Micromotives and Macrobehavior*.

page 260. A. Smith, *The Theory of Moral Sentiments* (1753; Oxford: Clarendon Press, 1976), pp. 124–125.

page 260. J. S. Mill, *On Utilitarianism* (1863).

page 261. A. Sen, Rational fools, *Philosophy and Public Affairs* 6 (1976–77): 317–344.

page 262. Sen, Rational fools, p. 329.

page 262. Sen, Rational fools, p. 336.

page 263. Hirsch, *Social Limits to Growth*.

page 267. Hirsch, *Social Limits to Growth*, p. 88.

page 267. Hirsch, *Social Limits to Growth*, p. 101.

page 268. MacIntyre, *After Virtue*.

page 270. J. D. Watson, *The Double Helix* (New York: Atheneum, 1968).

page 274. MacIntyre, *After Virtue*, pp. 175–176.

page 280. K. Marx, *The Poverty of Philosophy* (London: Lawrence and Wishart, 1955), p. 29.

Chapter 10

The ideas discussed in this chapter have been distilled principally from

F. Hirsch, *Social Limits to Growth* (Cambridge: Harvard University Press, 1976), and from M. Walzer, *Spheres of Justice* (New York: Basic Books, 1983).

page 281. W. S. Vickrey, An exchange of questions between economics and philosophy. In *Economic Justice,* ed. E. S. Phelps (Hammondsworth: Penguin, 1973), p. 60; J. K. Galbraith, *The New Industrial State* (New York: Signet, 1960), p. 404.

page 285. Aristotle, *Nicomachean Ethics.*

page 286. Hirsch, *Social Limits to Growth,* p. 150.

page 288. Walzer, *Spheres of Justice.*

page 291. Hirsch, *Social Limits to Growth,* p. 118.

page 292. Walzer, *Spheres of Justice,* pp. 82–83.

page 293. K. Arrow, Gifts and exchanges, *Philosophy and Public Affairs* 1 (1972): 357.

page 294. New York *Times,* August 19, 1984, pp. 3–1, 3–12, 3–13.

page 294. New York *Times,* June 9, 1985, pp. 3–1, 3–6.

page 295. M. Green and J. F. Berry, *The Challenge of Hidden Profits: Reducing Corporate Bureaucracy and Waste* (New York: Morrow, 1985).

page 295. Hirsch, *Social Limits to Growth,* pp. 117–118.

page 295. L. Robbins, *The Theory of Economic Policy in Classical Economics* (London: Macmillan, 1952), p. 177.

page 296. A. C. Pigou, *Economics of Welfare,* 3rd ed. (London: Macmillan, 1929), p. 11.

page 298. J. Schumpeter, *Capitalism, Socialism, and Democracy* (London: Unwin University Books, 1942).

page 299. D. C. Bennett and K. E. Sharpe, Is there a democracy "overload"? *Dissent* (Summer 1984).

page 299. S. P. Huntington, The United States. In *The Crisis of Democracy,* ed. M. Crozier, S. P. Huntington, and J. Watanuki (New York: New York University Press, 1975), p. 114.

page 301. A. Hirschman, *The Passions and the Interests* (Princeton: Princeton University Press, 1977).

page 301. Walzer, *Spheres of Justice.*

page 303. I am indebted to Elizabeth Anderson for this discussion of the characteristics that market goods have in common.

page 304. B. Pascal, *Pensées,* trans. J. M. Cohen (1670; Hammondsworth: Penguin, 1961), p. 96.

Chapter 11

page 310. K. Vonnegut, *Cat's Cradle* (New York: Holt, Rinehart and Winston, 1961), p. 227.

Epilogue

page 323. M. Sahlins, *The Use and Abuse of Biology* (Ann Arbor; University of Michigan Press, 1976), p. 100.

page 325. C. Geertz, *New York Review of Books,* January 24, 1980. p. 4.

Index